Methods in Molecular Biology

Series Editor
John M. Walker
School of Life and Medical Sciences
University of Hertfordshire
Hatfield, Hertfordshire, AL10 9AB, UK

For further volumes:
http://www.springer.com/series/7651

Diagnostic Bacteriology

Methods and Protocols

Edited by

Kimberly A. Bishop-Lilly

Germantown, MD, USA

Humana Press

Editor
Kimberly A. Bishop-Lilly
Germantown, MD, USA

ISSN 1064-3745 ISSN 1940-6029 (electronic)
Methods in Molecular Biology
ISBN 978-1-4939-7035-3 ISBN 978-1-4939-7037-7 (eBook)
DOI 10.1007/978-1-4939-7037-7

Library of Congress Control Number: 2017939131

Cover illustration: Cover image credit to Heidi Bishop.

Printed on acid-free paper

This Humana Press imprint is published by Springer Nature
The registered company is Springer Science+Business Media LLC
The registered company address is: 233 Spring Street, New York, NY 10013, U.S.A.

Preface

This volume in the Methods in Molecular Biology series is a comprehensive collection of protocols in molecular diagnostics of bacteria that will suit the needs of molecular biologists, clinical laboratorians, and physician scientists alike. The benefits of applying a molecular diagnostic approach for bacterial detection and identification are many—including reduced time (especially for difficult to culture or slow-growing organisms) and, in some cases, higher specificity (such as strain-level identification rather than species-level identification). In the case of high-throughput sequencing, one may even collect all the information one needs to know about a pathogen (such as genus-, species-, and strain-level identification along with virulence potential and antibiotic resistance potential) from a single reaction, rather than by applying several different highly specialized culture-based assays.

With protocols that are specific for common bacterial pathogens as well as protocols that can be applied to diverse or even unknown pathogens, this volume is a valuable resource for anyone who wishes to delve into the molecular age of diagnostics. Topics included range from duplex real-time PCR to next-generation sequencing and the associated bioinformatic analyses. The protocols contained within this book truly represent a range of assay types that are all on the cutting edge of diagnostic bacteriology.

Germantown, MD, USA *Kimberly A. Bishop-Lilly*

Contents

Contributors

DAN I. ANDERSSON • *Department of Medical Biochemistry and Microbiology, Uppsala University, Uppsala, Sweden*

JEAN ARMENGAUD • *CEA-Marcoule, DRF/JOLIOT/DMTS/SPI/Li2D, Laboratory "Innovative Technologies for Detection and Diagnostics", Bagnols-sur-Cèze, France*

NUNO F. AZEVEDO • *LEPABE, Laboratory for Process Engineering, Environment, Biotechnology and Energy, Department of Chemical Engineering, Faculty of Engineering, University of Porto, Porto, Portugal*

IVAN BARIŠIĆ • *Molecular Diagnostics, AIT Austrian Institute of Technology, Vienna, Austria*

SOPHIE BERTRAND • *National Reference Centre for Salmonella and Shigella, Bacterial Diseases Division, Communicable and Infectious Diseases, Scientific Institute of Public Health (WIV-ISP), Brussels, Belgium*

RENÉ BROUILLET • *Institute of Microbiology, University Hospital Center, University of Lausanne, Lausanne, Switzerland*

AMANDA CLAIRE BROWN • *Oxford Gene Technology, Oxford, UK; Department of Microbiology and Immunology, Cornell University, Ithaca, NY, USA*

NUNO CERCA • *CEB—Centre of Biological Engineering, LIBRO—Laboratory of Research in Biofilms Rosário Oliveira, University of Minho, Braga, Portugal*

KAMFAI CHAN • *AI Biosciences, Inc., College Station, TX, USA*

METTE T. CHRISTIANSEN • *University College London, London, UK*

KATHERINE E. CILWA • *Department of Regenerative Medicine, Naval Medical Research Center, Silver Spring, MD, USA*

PAUL COS • *Laboratory of Microbiology, Parasitology and Hygiene (LMPH), Faculty of Pharmaceutical, Biomedical and Veterinary Sciences, University of Antwerp, Antwerp, Belgium*

NICOLE J. CRANE • *The Department of Surgery at Uniformed Services University of the Health Sciences & The Walter Reed National Military Medical Center, Bethesda, MD, USA*

SILVIO L. CRAVERO • *Institute of Biotechnology, Center for Research in Veterinary and Agronomic Sciences - National Institute of Agricultural Technology (INTA), Hurlingham, Buenos Aires, Argentina*

SIGRID C.J. DE KEERSMAECKER • *Platform Biotechnology and Molecular Biology, Scientific Institute of Public Health (WIV-ISP), Brussels, Belgium*

TIM J. DUMONCEAUX • *Agriculture and Agri-Food Canada Saskatoon Research Centre, Saskatoon, SK, Canada; Department of Veterinary Microbiology, University of Saskatchewan, Saskatoon, SK, Canada*

ERIC A. ELSTER • *The Department of Surgery atUniformed Services University of the Health Sciences & The Walter Reed National Military Medical Center, Bethesda, MD, USA*

CEU FIGUEIREDO • *i3s Instituto de Investigação e Inovação em Saúde, Universidade do Porto, Porto, Portugal; IPATIMUP, Institute of Molecular Pathology and Immunology of the University of Porto, Porto, Portugal; FMUP, Faculty of Medicine of the University of Porto, Porto, Portugal*

SILVIA FONTENETE • *LEPABE, Laboratory for Process Engineering, Environment, Biotechnology and Energy, Department of Chemical Engineering, Faculty of Engineering, University of Porto, Porto, Portugal; i3s Instituto de Investigação e Inovação em Saúde, Universidade do Porto, Porto, Portugal; IPATIMUP, Institute of Molecular Pathology and Immunology of the University of Porto, Porto, Portugal; Nucleic Acid Center, Department of Physics, Chemistry and Pharmacy, University of Southern Denmark, Odense M, Denmark; ICBAS, Institute of Biomedical Sciences Abel Salazar, University of Porto, Porto, Portugal; Cancer Cell Biology Program, Epithelial Cell Biology Group, Spanish National Cancer Research Centre, Madrid, Spain*

STEPHEN FORSYTHE • *Pathogen Research Group, School of Science & Technology, Nottingham Trent University, Nottingham, UK*

MERON GHEBREMEDHIN • *Department of Regenerative Medicine, Naval Medical Research Center, Silver Spring, MD, USA*

STEPHEN H. GILLESPIE • *School of Medicine, University of St. Andrews, St. Andrews, UK*

ANDREA K. GIOFFRÉ • *Institute of Biotechnology, Center for Research in Veterinary and Agronomic Sciences - National Institute of Agricultural Technology (INTA), Hurlingham, Buenos Aires, Argentina; National Scientific and Technical Research Council (CONICET), Buenos Aires, Argentina*

GILBERT GREUB • *Institute of Microbiology, University Hospital Center, University of Lausanne, Lausanne, Switzerland*

CHRISTINE HAMMOND • *Agriculture and Agri-Food Canada Saskatoon Research Centre, Saskatoon, SK, Canada*

KATIA JATON • *Institute of Microbiology, University Hospital Center, University of Lausanne, Lausanne, Switzerland*

SUSAN JOSEPH • *Institute of Dentistry, Barts and The London School of Medicine & Dentistry, Queen Mary University of London, London, UK*

JENNIFER M. KOFONOW • *Division of Infectious Disease, Department of Medicine, University of Colorado, Aurora, CO, USA*

MARINA LEITE • *i3s Instituto de Investigação e Inovação em Saúde, Universidade do Porto, Porto, Portugal; IPATIMUP, Institute of Molecular Pathology and Immunology of the University of Porto, Porto, Portugal*

ANTONIO MACHADO • *Microbiology Institute, Universidad San Francisco de Quito, Cumbayá, Quito, Ecuador*

KATHLEEN MARCHAL • *Department of Plant Biotechnology and Bioinformatics, Ghent University, Ghent, Belgium; Department of Information Technology, Ghent University, IMinds, Ghent, Belgium*

SALVATORE A.E. MARRAS • *Department of Microbiology, Biochemistry and Molecular Genetics, Rutgers-New Jersey Medical School, Newark, NJ, USA*

WESLEY MATTHEUS • *National Reference Centre for Salmonella and Shigella, Bacterial Diseases Division, Communicable and Infectious Diseases, Scientific Institute of Public Health (WIV-ISP), Brussels, Belgium*

ANJA MEZGER • *Broad Institute of MIT and Harvard, Cambridge, MA, USA; Department of Biological Engineering, Massachusetts Institute of Technology, Cambridge, MA, USA*

MATS NILSSON • *Science for Life Laboratory, Department of Biochemistry and Biophysics, Stockholm University, Solna, Sweden*

Charles E. Okolie • *Department of Biological Sciences, Landmark University, Omu-Aran, Kwara State, Nigeria*
Chrystel Olivier • *Agriculture and Agri-Food Canada Saskatoon Research Centre, Saskatoon, SK, Canada*
Onya Opota • *Institute of Microbiology, University Hospital Center, University of Lausanne, Lausanne, Switzerland*
Katarina Oravcova • *School of Medicine, University of St. Andrews, St. Andrews, UK*
Nikhat Parveen • *Department of Microbiology, Biochemistry and Molecular Genetics, Rutgers-New Jersey Medical School, Newark, NJ, USA*
Rebecca L. Pavlicek • *Naval Medical Research Center-Asia, Singapore, Singapore*
Edel Pérez-López • *Instituto de Biotecnología y Ecología Aplicada (INBIOTECA), Universidad Veracruzana, Xalapa, Veracruz, Mexico*
Valeria Liliana Herrera Sevilla • *Universidad de las Fuerzas Armadas ESPE, Sangolquí, Ecuador*
Vijay R. Ramakrishnan • *Department of Otolaryngology-Head and Neck Surgery, University of Colorado, Aurora, CO, USA*
Nancy H.C. Roosens • *Platform Biotechnology and Molecular Biology, Scientific Institute of Public Health (WIV-ISP), Brussels, Belgium*
Wilber Sabiiti • *School of Medicine, University of St. Andrews, St. Andrews, UK*
Samantha Schlachter • *Department of Microbiology, Biochemistry and Molecular Genetics, Rutgers-New Jersey Medical School, Newark, NJ, USA*
Marcos D. Trangoni • *Institute of Biotechnology, Center for Research in Veterinary and Agronomic Sciences - National Institute of Agricultural Technology (INTA), Hurlingham, Buenos Aires, Argentina*
Thad W. Vickery • *University of Colorado School of Medicine, Aurora, CO, USA*
Maria Fernanda Loayza Villa • *Instituto de Microbiologia, Universidad San Francisco de Quito, Quito, Ecuador; Universidad de las Fuerzas Armadas ESPE, Sangolquí-Ecuador*
Nicolás Vivar-Díaz • *Laboratorios NETLAB S.A., Quito, Ecuador*
Noa Wolff • *Molecular Diagnostics, AIT Austrian Institute of Technology, Vienna, Austria*
Véronique Wuyts • *Platform Biotechnology and Molecular Biology, Scientific Institute of Public Health (WIV-ISP), Brussels, Belgium*

Chapter 1

Whole-Genome Enrichment Using RNA Probes and Sequencing of *Chlamydia trachomatis* Directly from Clinical Samples

Amanda Claire Brown and Mette T. Christiansen

Abstract

Whole-genome sequencing is a powerful, high-resolution tool that can be used to generate accurate data on bacterial population structure, phylogeography, and mutations associated with antimicrobial resistance. The ability to sequence pathogen genomes directly from clinical specimens, without the requirement for in vitro culturing, is attractive in terms of time- and labor-saving, especially in the case of slow-growing, or obligate intracellular pathogens, such as *Chlamydia trachomatis.* However clinical samples typically contain too low levels of pathogen nucleic acid, plus relatively high levels of human and natural microbiota DNA/RNA, to make this a viable option. Using a combination of whole-genome enrichment and deep sequencing, which has been proven to be a non-mutagenic approach, we can capture all known variations found within *C. trachomatis* genomes. The method is a consistent and sensitive tool that enables rapid whole-genome sequencing of *C. trachomatis* directly from clinical samples and has the potential to be adapted to other pathogens with a similar clonal nature.

Key words Whole-genome enrichment, Whole-genome sequencing, *Chlamydia trachomatis*, Clinical samples, SureSelect^{XT}

1 Introduction

The introduction of relatively affordable, desktop next generation sequencing (NGS) machines, such as Illumina's MiSeq and NextSeq, and Life Technologies' Ion PGM and Ion Torrent, has revolutionized the ability to perform whole genome sequencing (WGS) on small sized genomes, such as those of bacterial and viral pathogens. It is now becoming possible to introduce WGS into diagnostic microbiology laboratories [1–3]. However, clinical samples typically contain too low levels of pathogen nucleic acid, plus relatively high levels of human DNA/RNA as well as DNA/RNA from the natural microbiota, to allow sequencing without an in vitro culture. Adding a culturing step can add significant delay to data acquisition, especially in the case of organisms that are

Kimberly A. Bishop-Lilly (ed.), *Diagnostic Bacteriology: Methods and Protocols*, Methods in Molecular Biology, vol. 1616, DOI 10.1007/978-1-4939-7037-7_1, © Springer Science+Business Media LLC 2017

difficult to culture, such as *Chlamydia trachomatis*, and organisms that are slow-growing, such as *Mycobacterium tuberculosis*.

C. trachomatis is the most common bacterial agent in sexually transmitted infections (STI), globally accounting for more than 100 million infections per year [4, 5]. *C. trachomatis* causes pelvic inflammatory disease in women, which has severe long-term consequences including ectopic pregnancy, infertility, and chronic pelvic pain; in addition serovars A–C cause the blinding disease, trachoma, which affects millions of people worldwide [6, 7]. The loss of productivity and morbidity resulting from *C. trachomatis* infection has been recognized as placing a heavy economic burden on society [8, 9].

C. trachomatis strains are classified into two biovars: the ocular/urogenital biovar and the lymphogranuloma venereum (LGV) biovar [10]. The two biovars can be subdivided into 15–19 different serovars. Further genotypic classification is based on nucleotide sequencing of the *ompA* gene, which encodes the major outer membrane protein and is the target of serotype classification. The ocular/urogenital biovar consists of the ocular serovars A–C and the urogenital serovars D–K, all of which are usually confined to mucosal epithelia whereas the LGV biovar, consisting of serotypes L1–L3, is more invasive and can disseminate to other tissues and the draining lymphatic system. It has previously been demonstrated that genotyping of the *ompA* gene is insufficient for exploring *C. trachomatis* population structure and performing molecular epidemiological studies on transmission, as this region undergoes high levels of recombination [11]. Also variation within the *ompA* gene differs among serovars and sexual networks can be predominated by a single serovar, making strain distribution and evolutionary studies impractical [12, 13]. In this context, whole-genome sequencing (WGS) has been used to generate accurate data on bacterial population structure and phylogeography [14]. In addition whole-genome sequencing can also facilitate the identification of mutations associated with antimicrobial resistance [3, 15].

Clinical samples often contain low numbers of pathogens and to obtain sufficient material for WGS of *C. trachomatis*, in vitro culture is usually required; however, as *C. trachomatis* is an obligate intracellular pathogen it is labor-intensive to grow in vitro [16]. For this reason methods that allow sequencing directly from *C. trachomatis* positive samples are particularly attractive. An antibody-based enrichment method, which targets intact *C. trachomatis* cells, followed by whole genome amplification of the total DNA within the sample was recently described [17]. This method proved useful for sequencing *C. trachomatis* from complex clinical samples but showed only a 15–30% success rate, which underlines the need for a more reliable methodology.

The method described in this chapter utilizes the SureSelect^{XT} Target-Enrichment protocol (Agilent Technologies), which uses custom designed 120-mer RNA oligonucleotides. These RNA "baits" span the entire genome, and can recover (by hybridization) low copy number pathogens from clinical samples with sufficiently high sensitivity and specificity to enable ultra-deep whole-genome sequencing. Using this method we were able to obtain full length *C. trachomatis* genomes (>95–100% coverage of a reference genome) from eight of ten clinical samples tested (seven vaginal swabs and three urine samples). We also tested nine cultured *C. trachomatis* samples, representing different serovars, and obtained full length genomes for all. The proportion of reads mapping to *C. trachomatis* and the mean read depth across each genome were strongly linked to the number of bacterial copies within the original sample. Phylogenetic analysis confirmed the known population structure and the data showed potential for identification of minority variants and mutations associated with antimicrobial resistance. The sensitivity of this method was tenfold higher than other reported methodologies, and offers the opportunity for gaining a wider understanding of *C. trachomatis* population structures, transmission patterns, and the evolution of antimicrobial resistance. We have successfully used this approach on not just *C. trachomatis* but also herpesviruses (including cytomegalovirus (CMV)), *M. tuberculosis* [18–21], hepatitis B, hepatitis C, human immunodeficiency virus (HIV), influenza A, and norovirus (publications currently in preparation).

2 Materials

In addition to the materials listed below sterile, nuclease-free, aerosol barrier pipette tips should be used for all liquid manipulations, and standard molecular biology protocol, such as the wearing of gloves, taking care to not cross-contaminate samples etc., should be performed.

2.1 DNA Extraction and Quantification

1. QIAmp Mini Kit (Cat. No. 51304), Qiagen (*see* **Note 1**).
2. DNA LoBind 1.5 mL tubes (022431021, Eppendorf).
3. PBS (sterile).
4. Vortex mixer.
5. 1.5 mL tube heat block set at 56 °C.
6. Microfuge.
7. 96–100% ethanol, molecular biology grade.
8. Nuclease-free water (P1193, Promega).
9. Qubit dsDNA HS Assay Kit (Q32854), and Qubit 3.0 Fluorometer (Q33216), Life Technologies.
10. Qubit Assay Tubes (Q32856), Life Technologies.
11. 50 mL sterile Falcon tubes (352070), Corning.

2.2 DNA Shearing

1. Covaris E210 Focused-ultrasonicator.
2. Covaris microTUBE with AFA fiber and snap cap (520045, Covaris).
3. TE_{LOW} buffer (10 mM Tris–HCl, pH 8.0, 0.1 mM EDTA, 12090-015, Life Technologies).
4. Human genomic DNA (G3041, Promega), (*see* **Note 2**).
5. DNA LoBind 1.5 mL tubes (022431021, Eppendorf).
6. Microfuge.

2.3 Post Shearing Sample Purification

1. Agencourt AMPureXP beads (A63880, Beckman Coulter).
2. DynaMag-2 magnetic stand (12321D, Life Technologies).
3. 70% ethanol (*see* **Note 3**).
4. Nuclease-free water (P1193, Promega).
5. DNA LoBind 1.5 mL tubes (022431021, Eppendorf).
6. Microfuge.
7. Heat block set at 37 °C.
8. 2200 TapeStation Nucleic Acid System (G2965AA, Agilent).
9. D1000 ScreenTape (5067-5582, Agilent) and D1000 reagents (5067-5583, Agilent).
10. Thin-wall 8-strip tubes (3148, Thermo Scientific) and caps (3148C, Thermo Scientific).

2.4 End Repair, A-Tailing, and Adapter Ligation

1. SureSelectXT Reagent Kit, MSQ, 16 (G9612A, Agilent) (*see* **Note 4**).
2. End Repair mix (52 μL per sample): 35.2 μL nuclease-free water, 10 μL 10× End Repair Buffer, 1.6 μL dNTP mix, 1 μL T4 DNA polymerase, 2 μL Klenow DNA polymerase, and 2.2 μL T4 Polynucleotide kinase.
3. A-Tailing Mix (20 μL per sample): 11 μL nuclease-free water, 5 μL 10× Klenow DNA polymerase buffer, 1 μL ATP, and 3 μL Exo (−) Klenow DNA polymerase.
4. Adapter Ligation Mix (37 μL per sample): 15.5 μL nuclease-free water, 10 μL 5× T4 DNA ligase buffer, 10 μL Adaptor Oligo mix, and 1.5 μL T4 DNA ligase.
5. Agencourt AMPureXP beads (A63880, Beckman Coulter).
6. DynaMag-96 Side magnet (12331D, Life Technologies).
7. 70% ethanol (*see* **Note 3**).
8. Nuclease-free water (P1193, Promega).
9. Thin-wall 8-strip tubes (3148, Thermo Scientific) and caps (3148C, Thermo Scientific).
10. 2200 TapeStation Nucleic Acid System (G2965AA, Agilent).

11. D1000 ScreenTape (5067-5582, Agilent) and D1000 reagents (5067-5583, Agilent).
12. Thermal cycler, e.g., SureCycler 8800 (G8800A, Agilent), with 96-well module (G8810A, Agilent).
13. Microfuge.
14. 5–50 μL 8-channel pipette (e.g., P4808-50, Labnet), and 20–200 μL 8-channel pipette (e.g., P4508-200, Labnet).

2.5 Amplification of Adapter-Ligated Library

1. SureSelectXT Reagent Kit, MSQ, 16 (G9612A, Agilent).
2. Herculase II Fusion DNA Polymerase (600675, Agilent).
3. Pre-Capture PCR Mix (35 μL per sample): 21 μL nuclease-free water, 1.25 μL SureSelect Primer (forward), 1.25 μL SureSelect Primer (forward), 10 μL Herculase II 5× reaction buffer, 0.5 μL dNTP (100 mM), and 1 μL Herculase II polymerase.
4. Agencourt AMPureXP beads (A63880, Beckman Coulter).
5. DynaMag-96 Side magnet (12331D, Life Technologies).
6. 70% ethanol (*see* **Note 3**).
7. Nuclease-free water (P1193, Promega).
8. Thin-wall 8-strip tubes (3148, Thermo Scientific) and caps (3148C, Thermo Scientific).
9. 2200 TapeStation Nucleic Acid System (G2965AA, Agilent).
10. D1000 ScreenTape (5067-5582, Agilent) and D1000 reagents (5067-5583, Agilent).
11. Thermal cycler, e.g., SureCycler 8800 (G8800A, Agilent), with 96-well module (G8810A, Agilent).
12. 5–50 μL 8-channel pipette (e.g., P4808-50, Labnet), and 20–200 μL 8-channel pipette (e.g., P4508-200, Labnet).

2.6 Hybridization of Adapter-Ligated Library to RNA Baits

1. SureSelectXT Reagent Kit, MSQ, 16 (G9612A, Agilent).
2. SureSelect Block Mix (5.6 μL per sample): 2.5 μL SureSelect block #1 (green cap), 2.5 μL SureSelect block #2 (blue cap), and 0.6 μL SureSelect block #3 (brown cap).
3. Hybridization Buffer (24.5 μL per sample - includes excess): 12.5 μL SureSelect Hyb #1 (orange cap), 0.5 μL SureSelect Hyb #2 (red cap), 5 μL SureSelect Hyb #3 (yellow cap), and 6.5 μL SureSelect Hyb #4 (black cap).
4. *C. trachomatis* RNA baits (Oxford Gene Technologies), *see* **Note 5**.
5. Vacuum concentrator (DNA SpeedVac 120, Thermo Scientific).
6. Thin-wall 96-well PCR Plates (3146, Thermo Scientific) and caps (3148C, ThermoScientific).

7. 5–50 μL 8-channel pipette (P4808-50, Labnet).
8. Thermal cycler, e.g., SureCycler 8800 (G8800A, Agilent), with 96-well module (G8810A, Agilent).
9. 1.5 mL tube heat block set at 65 °C.
10. Dynabeads MyOne Streptavidin T1 (65601, Life Technologies).
11. DynaMag-2 magnetic stand (12321D, Life Technologies).
12. Nutator (421105, BD).
13. Nuclease-free water (P1193, Promega).
14. Vortex mixer.
15. Microfuge.
16. DNA LoBind 1.5 mL tubes (022431021, Eppendorf).

2.7 Addition of Index Tags by Post-hybridization Amplification

1. SureSelectXT Reagent Kit, MSQ, 16 (G9612A, Agilent).
2. Herculase II Fusion DNA Polymerase (600675, Agilent).
3. Post-Capture PCR mix (34 μL per sample): 21.5 μL nuclease-free water, 1 μL SureSelect Indexing Post Capture PCR (forward) primer (orange cap), 10 μL Herculase II 5× reaction buffer, 0.5 μL dNTP (100 mM), and 1 μL Herculase II polymerase.
4. Agencourt AMPureXP beads (A63880, Beckman Coulter).
5. DynaMag-96 Side magnet (12331D, Life Technologies).
6. 70% ethanol (*see* **Note 3**).
7. Nuclease-free water (P1193, Promega).
8. Thin-wall 8-strip tubes (3148, ThermoScientific) and caps (3148C, ThermoScientific).
9. 2200 TapeStation Nucleic Acid System (G2965AA, Agilent).
10. D1000 ScreenTape (5067-5582, Agilent) and D1000 reagents (5067-5583, Agilent).
11. Thermal cycler, e.g., SureCycler 8800 (G8800A, Agilent), with 96-well module (G8810A, Agilent).
12. 5–50 μL 8-channel pipette (e.g., P4808-50, Labnet), and 20–200 μL 8-channel pipette (e.g., P4508-200, Labnet).

2.8 Pooling Samples for Multiplexed Sequencing

1. Nuclease-free water (P1193, Promega).
2. DNA LoBind 1.5 mL tubes (022431021, Eppendorf).
3. Vortex mixer.

2.9 Illumina Sequencing

1. MiSeq desktop sequencer (Illumina).
2. MiSeq Reagent Kit v2 (300 cycle, MS-102-2002, Illumina).
3. 10 M NaOH (72068-100 mL, Sigma) (*see* **Note 6**).
4. PhiX Control (FC-110-3001, Illumina).

5. 10 mM Tris–HCl, pH 8.5 with 0.1% Tween 20.
6. Nuclease-free water (P1193, Promega).
7. Vortex mixer.

2.10 Data Analysis

1. CLC Genomics Workbench (version 6.5.0/6.5.1) including the CLC Microbial Genome Finishing Module (version 1.2.1/1.3.0) (Qiagen).

3 Methods

3.1 DNA Extraction and Quantification

1. Using the QIAamp DNA extraction kit, pipette 20 μL Qiagen Protease into the bottom of a 1.5 mL DNA LoBind tube.
2. For vaginal swabs suspend cells in 200 μL sterile PBS and add to the 1.5 mL tube, for urine add 200 μL; if the sample volume is lower than 200 μL make up to 200 μL with PBS.
3. Add 200 μL Buffer AL and mix by pulse vortexing for 15 s. Incubate at 56 °C for 10 min. Briefly spin down tube to collect sample.
4. Add 200 μL ethanol (96–100%) to the sample, mix by pulse-vortexing for 15 s, and briefly centrifuge to collect.
5. Insert a QIAamp Mini spin column into a 2 mL collection tube, carefully add the sample to the column without wetting the rim. Close the cap and centrifuge at 6000 × *g* for 1 min. Remove QIAamp column to a fresh 2 mL collection tube, discard the tube containing the flow-through (*see* **Note** 7).
6. Add 500 μL Buffer AW1 to the column, spin at 6000 × *g* for 1 min, remove QIAamp column to a fresh 2 mL collection tube, discard the tube containing the flow-through (*see* **Note** 7).
7. Add 500 μL Buffer AW2 to the column, spin at full speed for 1 min, remove QIAamp column to a fresh 2 mL collection tube, and spin at top speed for 1 min; discard the tube containing the flow-through.
8. Place QIAamp column in a new DNA LoBind tube, add 100 μL Buffer AE and incubate at room temperature for 5 min, then spin at 6000 × *g* for 1 min.
9. Label lids of Qubit assay tubes, number of sample plus 2 standards. Prepare Qubit working solution by diluting Qubit dsDNA HS reagent 1:200 with Qubit dsDNA HS Buffer in a 50 mL plastic tube, allow for 200 μL/sample or standard.
10. Add 190 μL of Qubit working solution to each standard tube, add 199 μL of Qubit working solution to each sample tube.
11. Add 10 μL of each standard to the appropriate tube, and 1 μL of each sample. Mix by vortexing for 2–3 min, take care not to

create bubbles. Briefly spin tubes down in a microfuge. Incubate at room temperature for 2 min.

12. Select "DNA" on the Qubit 3.0 Fluorometer, and then "ds High Sensitivity" as the assay. Press "Read Standards" to proceed; insert tube containing standard #1 and press "Read standard," remove tube when read is complete and insert tube containing standard #2, press "Read standard." When complete press "Run samples," select the sample volume (1 μL) and units (ng/μL), insert a sample tube and press "Read tube," remove tube when reading is complete and insert next sample, until all are read.

3.2 DNA Shearing

1. Dilute samples to 3 μg DNA/130 μL using TE_{LOW} in a DNA LoBind tube, where total DNA concentration is lower than 3 μg use human DNA (Promega) to bulk to a final concentration of 3 μg (*see* **Note 2**).
2. Using a pipette tip introduce 130 μL DNA sample to Covaris microTube directly through the pre-split septa, do not introduce bubbles.
3. Secure the microTube in the tube holder and shear for 6 × 60 s (duty cycle 10%, intensity 5, 200 cycles per burst with frequency sweeping) (*see* **Note 8**).
4. Remove microTube from Covaris, spin briefly in microfuge.
5. Remove sample, using a pipette tip, and transfer into a new DNA LoBind tube.

3.3 Post Shearing Sample Purification

1. Add 180 μL of homogenous AMPureXP beads (*see* **Note 9**) to each sheared sample tube (~130 μL), mix by vortexing for 5 s, incubate at room temperature for 5 min.
2. Briefly spin to collect sample to bottom of the tube.
3. Place tube on magnetic stand and allow the solution to clear (3–5 min).
4. Keeping the tube on the magnetic stand, carefully remove the cleared supernatant without disturbing the beads.
5. With the tube still in place on the magnetic stand add 500 μL 70% ethanol, incubate for 1 min or when solution clears. Remove ethanol wash and repeat.
6. After second wash step use a P10 tip to remove any residual ethanol from the bottom of the tube.
7. Dry samples on heat block at 37 °C for no more than 5 min (*see* **Note 10**).
8. Add 52 μL nuclease-free water directly to bead pellet. Mix by vortexing and briefly spin tube to collect contents. Incubate for 2 min at room temperature.

9. Put tube on magnetic rack, leave for 2–3 min to clear and collect the supernatant (~50 μL), transfer to a new DNA LoBind tube.
10. Add 3 μL D1000 Sample Buffer and 1 μL D1000 ladder to tube 1 using an 8-tube strip (*see* **Note 11**).
11. Add 3 μL D1000 Sample Buffer and 1 μL cleaned, sheared sample to tubes 2–16 using 8-tube strips (*see* **Note 12**), add caps and vortex, spin down for 1 min.
12. Load samples into TapeStation, and run D1000 tape. Assess quality and quantity, with a peak height between 150 and 200 nucleotides.
13. Samples can be stored overnight in the fridge at 4 °C, or at −20 °C for longer terms.

3.4 End Repair, A-Tailing, and Adapter Ligation

1. Prepare End Repair Mix for each sample, on ice.
2. Pipette 52 μL End Repair Mix into thin-wall 8-strip tubes, add 48 μL of each sheared and cleaned DNA sample, mix by gentle pipetting, and close strip caps.
3. Incubate on a thermocycler at 20 °C for 30 min; do not use a heated lid.
4. Add 180 μL of homogenous AMPureXP beads (*see* **Note 9**) to each sheared sample tube (~100 μL), mix by pipetting until homogenous, incubate at room temperature for 5 min.
5. Place strip tubes on DynaMag-96 Side magnet and allow the solution to clear (3–5 min).
6. Keeping the tube on the magnetic stand, carefully remove the cleared supernatant without disturbing the beads.
7. With the tube still in place on the magnetic stand add 200 μL 70% ethanol, incubate for 1 min or when solution clears. Remove ethanol wash and repeat.
8. After second wash step use a P10 tip to remove any residual ethanol from the bottom of the tube.
9. Dry samples on thermocycler at 37 °C for no more than 5 min (*see* **Note 10**).
10. Add 32 μL nuclease-free water directly to bead pellet. Mix by vortexing and briefly spin tube to collect contents. Incubate for 2 min at room temperature.
11. Put tube on magnetic rack, leave for 2–3 min to clear and collect the supernatant (~30 μL), transfer to a new 8 strip tube.
12. Samples can be stored overnight in the fridge at 4 °C, or at −20 °C for longer terms.
13. Prepare A-Tailing Mix for each sample, on ice.

14. Add 20 μL A-Tailing Mix to each 30 μL End Repaired Sample in 8 strip tube, mix by gentle pipetting, and close strip caps.
15. Incubate on a thermocycler at 37 °C for 30 min; do not use a heated lid.
16. Add 90 μL of homogenous AMPureXP beads (*see* **Note 9**) to each sheared sample tube (~50 μL), mix by pipetting until homogenous, incubate at room temperature for 5 min.
17. Place strip tubes on DynaMag-96 Side magnet and allow the solution to clear (3–5 min).
18. Keeping the tube on the magnetic stand, carefully remove the cleared supernatant without disturbing the beads.
19. With the tube still in place on the magnetic stand add 200 μL 70% ethanol, incubate for 1 min or when solution clears. Remove ethanol wash and repeat.
20. After second wash step use a P10 tip to remove any residual ethanol from the bottom of the tube.
21. Dry samples on thermocycler at 37 °C for no more than 5 min (*see* **Note 10**).
22. Add 15 μL nuclease-free water directly to bead pellet. Mix by vortexing and briefly spin tube to collect contents. Incubate for 2 min at room temperature.
23. Put tube on magnetic rack, leave for 2–3 min to clear and collect the supernatant (~13 μL), transfer to a new 8 strip tube.
24. Proceed immediately to adaptor ligation step.
25. Prepare Adaptor Ligation Mix for each sample, on ice.
26. Add 37 μL Adaptor Ligation Mix to each 13 μL A-Tailed Sample in a 8-tube strip, mix by gentle pipetting, and close strip caps.
27. Incubate on a thermocycler at 20C for 15 min, do not use a heated lid.
28. Add 90 μL of homogenous AMPureXP beads (*see* **Note 9**) to each sheared sample tube (~50 μL), mix by pipetting until homogenous, incubate at room temperature for 5 min.
29. Place strip tubes on DynaMag-96 Side magnet and allow the solution to clear (3–5 min).
30. Keeping the tube on the magnetic stand, carefully remove the cleared supernatant without disturbing the beads.
31. With the tube still in place on the magnetic stand add 200 μL 70% ethanol, incubate for 1 min or when solution clears. Remove ethanol wash and repeat.
32. After second wash step use a P10 tip to remove any residual ethanol from the bottom of the tube.

33. Dry samples on thermocycler at 37 °C for no more than 5 min (*see* **Note 10**).
34. Add 47 μL nuclease-free water directly to bead pellet. Mix by vortexing and briefly spin tube to collect contents. Incubate for 2 min at room temperature.
35. Put tubes on magnetic rack, leave for 2–3 min to clear and collect the supernatant (~45 μL), transfer to a new 8-tube strip.
36. Add 3 μL D1000 Sample Buffer and 1 μL D1000 ladder to tube 1 using an 8-tube strip (*see* **Note 11**).
37. Add 3 μL D1000 Sample Buffer and 1 μL sample to tubes 2–16 using 8-tube strips (*see* **Note 12**), add caps and vortex, spin down for 1 min, carefully remove caps.
38. Load samples into TapeStation, and run D1000 tape. Assess quality and quantity; the peak height should have increased to ~250–275 bp.
39. Samples can be stored overnight in the fridge at 4 °C, or at −20 °C for longer terms.

3.5 Amplification of Adapter-Ligated Library

1. Adjust, if required, adaptor ligated samples to 20 ng/μL using nuclease-free water.
2. Prepare Pre-Capture PCR Mix.
3. Add 35 μL Pre-Capture PCR Mix to each 15 μL Adaptor Ligated Sample in 8 strip tube, mix by gentle pipetting, and close strip caps.
4. Cycle on a thermocycler using the parameters in Table 1 below, repeating **steps 2–4** for 6–12 cycles (*see* **Note 13**).
5. Add 90 μL of homogenous AMPureXP beads (*see* **Note 9**) to each sheared sample tube (~50 μL), mix by pipetting until homogenous, incubate at room temperature for 5 min.
6. Place strip tubes on DynaMag-96 Side magnet and allow the solution to clear (3–5 min).

Table 1
Thermocycler conditions for amplification of adapter-ligated library

Step	Temperature (°C)	Time
1. Initial denaturing	98	2 min
2. Denaturing	98	30 s
3. Annealing	65	30 s
4. Extension	72	1 min
5. Final extension	72	10 min
6. Cool down	4	Hold

7. Keeping the tubes on the magnetic stand, carefully remove the cleared supernatant without disturbing the beads.
8. With the tube still in place on the magnetic stand add 200 μL 70% ethanol, incubate for 1 min or when solution clears. Remove ethanol wash and repeat.
9. After second wash step use a P10 tip to remove any residual ethanol from the bottom of the tube.
10. Dry samples on thermocycler at 37 °C for no more than 5 min (*see* **Note 10**).
11. Add 32 μL nuclease-free water directly to bead pellet. Mix by vortexing and briefly spin tube to collect contents. Incubate for 2 min at room temperature.
12. Put tubes on magnetic rack, leave for 2–3 min to clear and collect the supernatant (~30 μL), transfer to new 8-tube strip.
13. Add 3 μL D1000 Sample Buffer and 1 μL D1000 ladder to tube 1 using an 8-tube strip (*see* **Note 11**).
14. Add 3 μL D1000 Sample Buffer and 1 μL sample to tubes 2–16 using 8-tube strips (*see* **Note 12**), add caps and vortex, spin down for 1 min, remove caps gently.
15. Load samples into TapeStation, and run D1000 tape. Assess quality and quantity; the peak height should be ~250–275 bp.
16. Samples can be stored overnight in the fridge at 4 °C, or at −20 °C for longer terms.

3.6 Hybridization of Adapter-Ligated Library to RNA Baits

1. Concentrate Pre-Capture PCR samples using vacuum concentrator as required so they reach 147–220 ng/μL in nuclease-free water (*see* **Note 14**).
2. Using a 96-well plate (PLATE A) add 500–750 ng in 3.4 μL of each sample into separate wells of column 2, (*see* **Note 15**).
3. Prepare sufficient SureSelect Block Mix for all samples being hybridized.
4. Add 5.6 μL SureSelect Block Mix to each sample well in column 2 of the 96-well plate.
5. Add strip caps to column 2 and seal well, carefully check each to ensure a complete seal.
6. Incubate plate(s) with samples in column 2 for 95 °C for 5 min, and then hold at 65 °C, using a thermocycler with a heated lid (105 °C).
7. Prepare sufficient Hybridization Buffer for the samples being hybridized.
8. Whilst maintaining PLATE A at 65 °C load 20 μL Hybridization Buffer into the required wells in column 1.
9. Add strip caps to column 1 and seal well, carefully check each to ensure a complete seal.

10. Close thermocycler lid and incubate at 65 °C for a least 5 min before proceeding to next step.
11. Prepare SureSelect RNA Capture Baits by diluting with 10% RNase Block, for each sample: 2 μL RNA baits (< 3.0 MB), 0.5 μL RNase Block, 4.5 μL nuclease-free water; keep on ice.
12. In a new 96-well plate (PLATE B), on ice, add SureSelect RNA Capture Baits (7 μL) to column 3.
13. Using a multichannel pipette transfer 7 μL SureSelect RNA Capture Baits from column 3 of PLATE B, to column 3 of PLATE A, still in place on the thermocycler at 65 °C.
14. Add strip caps to column 3 of PLATE A and seal well, carefully check each to ensure a complete seal.
15. Close thermocycler lid and incubate at 65 °C for 2 min.
16. Whilst maintaining PLATE A at 65 °C, transfer 13 μL Hybridization Buffer from column 1 to column 3, using a multichannel pipette (*see* **Note 16**).
17. Whilst continuing to maintain PLATE A at 65 °C, transfer the entire contents of column 2 to column 3, using a multichannel pipette, still set at 13 μL (*see* **Note 16**).
18. Mix column 3 well by careful pipetting 8–10 times.
19. Add new strip caps to column 3 of PLATE A and seal well, carefully check each lip to ensure a complete seal (*see* **Note 17**).
20. Add strip caps to columns 1, 2, and 12 of PLATE A, to ensure the plate is balanced under the heated lid.
21. Continue to incubate the hybridization mixture, on the thermocycler for 16–24 h at 65 °C.
22. Prepare 50 μL Dynabeads MyOne Streptavidin T1 beads for each sample hybridized, add to a DNA LoBind tube and place on magnetic rack for 1–2 min.
23. Remove the cleared supernatant.
24. Add 200 μL SureSelect Binding Buffer.
25. Vortex for 5 s, spin down to collect tube contents and return to magnetic rack for 1–2 min.
26. Repeat **steps 23–25** for a total of three washes.
27. Resuspend beads in 200 μL SureSelect Binding Buffer and label each LoBind tube with sample identifier.
28. Open the lid to the thermocycler containing PLATE A, keeping the plate at 65 °C, open the first cap for column 3 (A3), remove the contents (typically 27–29 μL, *see* **Note 18**) and place in corresponding LoBind tube containing resuspended MyOne Streptavidin T1 beads. Invert 3–5 times to mix. Repeat for other samples (B3, C3, …).

29. Mix all tubes containing hybridized samples and MyOne Streptavidin T1 beads well on a vortex for 5 s, then transfer to a Nutator. Incubate on the Nutator for 30 min at room temperature (*see* **Note 19**).
30. Briefly spin the tubes in a microfuge to collect the contents, and then add to a magnetic rack, and allow to clear for 1–3 min.
31. Remove the cleared supernatant and discard.
32. Add 500 μL SureSelect Wash Buffer #1 to each tube, close caps and vortex for 5 s.
33. Incubate at room temperature for 15 min, mixing every 5 min using a vortex.
34. Briefly spin the tubes in a microfuge to collect the contents, and then add to a magnetic rack, and allow to clear for 1–3 min.
35. Remove the cleared supernatant and discard.
36. Add 500 μL pre-warmed 65 °C SureSelect Wash Buffer #2 (*see* **Note 20**) to each tube, close caps and vortex for 5 s.
37. Incubate on a heat block at 65 °C for 10 min, mixing every 3 min using a vortex.
38. Repeat **steps 34–37** for a total of 3 SureSelect Wash Buffer #2 washes.
39. After the final wash briefly spin tube before returning to the magnetic rack and use a P10 tip to ensure all residual wash buffer is removed.
40. Resuspend bead pellet in 45 μL nuclease-free water, leaving beads remaining in tube.
41. Samples should be stored at 4 °C.

3.7 Addition of Index Tags by Post-hybridization Amplification

1. Prepare Post-Capture PCR Mix.
2. Add 34 μL Post-Capture PCR Mix to required number of tubes in 8-strip tube.
3. Resuspend beads in hybridized samples, pipette 15 μL of bead mix into tube containing Post-Capture PCR Mix.
4. Add 1 μL of the appropriate PCR Index Primer to each reaction tube (*see* **Note 21**), mix by gentle pipetting, and close strip caps.
5. Cycle on a thermocycler using the parameters in Table 2 below, repeating **steps 2–4** for 12–18 cycles (*see* **Note 22**).
6. Add 90 μL of homogenous AMPureXP beads (*see* **Note 9**) to each sheared sample tube (~50 μL), mix by pipetting until homogenous, incubate at room temperature for 5 min.
7. Place strip tubes on DynaMag-96 Side magnet and allow the solution to clear (3–5 min).

Table 2
Thermocycler conditions for addition of index tags post-hybridization

Step	Temperature (°C)	Time
1. Initial denaturing	98	2 min
2. Denaturing	98	30 s
3. Annealing	57	30 s
4. Extension	72	1 min
5. Final extension	72	10 min
6. Cool down	4	Hold

8. Keeping the tubes on the magnetic stand, carefully remove the cleared supernatant without disturbing the beads.
9. With the tube still in place on the magnetic stand add 200 μL 70% ethanol, incubate for 1 min or when solution clears. Remove ethanol wash and repeat.
10. After second wash step use a P10 tip to remove any residual ethanol from the bottom of the tube.
11. Dry samples on thermocycler at 37 °C for no more than 5 min (*see* **Note 10**).
12. Add 32 μL nuclease-free water directly to bead pellet. Mix by vortexing and briefly spin tube to collect contents. Incubate for 2 min at room temperature.
13. Put tubes on magnetic rack, leave for 2–3 min to clear and collect the supernatant (~30 μL), transfer to new 8-tube strip.
14. Add 3 μL D1000 Sample Buffer and 1 μL D1000 ladder to tube 1 using an 8-tube strip (*see* **Note 11**).
15. Add 3 μL D1000 Sample Buffer and 1 μL sample to tubes 2–16 using 8-tube strips (*see* **Note 12**), add caps and vortex, spin down for 1 min, carefully remove caps.
16. Load samples into TapeStation, run D1000 tape. Assess quality and quantity, recording the peak height (*see* **Note 23**).
17. Label lids of Qubit assay tubes, number of sample plus 2 standards. Prepare Qubit working solution by diluting Qubit dsDNA HS reagent 1:200 with Qubit dsDNA HS Buffer in a 50 mL plastic tube; allow for 200 μL/sample or standard.
18. Add 190 μL of Qubit working solution to each standard tube; add 199 μL of Qubit working solution to each sample tube.
19. Add 10 μL of each standard to the appropriate tube, and 1 μL of each sample. Mix by vortexing for 2–3 min, taking care not to create bubbles. Briefly spin tubes down in a microfuge. Incubate at room temperature for 2 min.

20. Select "DNA" on the Qubit 3.0 Fluorometer, and then "ds High Sensitivity" as the assay. Press "Read Standards" to proceed; insert tube containing standard #1 and press "Read standard," remove tube when read is complete and insert tube containing standard #2, press "Read standard." When complete press "Run samples," select the sample volume (1 μL) and units (ng/μL), insert a sample tube and press "Read tube," remove tube when reading is complete and insert next sample, until all are read.
21. Samples can be stored overnight in the fridge at 4 °C, or at −20 °C for longer terms.

3.8 Pooling Samples for Multiplexed Sequencing

1. Use the concentration as determined by Qubit, and the size, as determined using TapeStation to calculate the molarity of each Post-Index Amplified sample (*see* **Notes 23** and **24**).
2. Combine the Post-Index Amplified samples so that each is representative in equimolar amounts in the library pool, using the formula below to determine the volume of each sample to use:

$$\textit{Volume of Post}-\textit{Index Amplified sample}=\frac{V(f)XC(f)}{\#XC(i)}$$

Where $V(f)$ is the final desired volume of the pool, $C(f)$ is the desired concentration of the pool (e.g., 2 nM for MiSeq v2), # is the total number of indexes in the pool, and $C(i)$ is the initial concentration of each Post-Index Amplified sample.

3. Mix pool well by vortexing, store at 4 °C.
4. Create a MiSeq sample sheet, using either Illumina Experiment Manager or Excel (*see* **Note 25**).

3.9 Illumina Sequencing

1. Thaw MiSeq Reagent Cartridge and HT1 buffer, overnight at 4 °C or at least 1 h at room temperature in a dish of cold water.
2. Once thawed keep Reagent Cartridge and HT1 buffer at 4 °C until needed.
3. Prepare a fresh dilution of 0.2 N NaOH (combine 980 μL of nuclease-free water and 20 μL 10 M NaOH in a LoBind tube).
4. Prepare PhiX control (*see* **Note 26**) by adding 2 μL 10 nM PhiX library with 3 μL 10 mM Tris–HCl, pH 8.5 with 0.1% Tween 20 in a LoBind tube. Mix well. Then add 5 μL 0.2 N NaOH, briefly vortex and then centrifuge at 280 × *g* for 1 min. Incubate for 5 min at room temperature, then add 990 μL prechilled HT1 buffer. This is 20pM denatured PhiX, for v2 reagents 12.5pM is required: combine 375 μL 20pM denatured PhiX with 225 μL prechilled HT1 (*see* **Notes 27** and **28**).
5. Denature the 5 μL of the 2 nM library pool by combining with 5 μL 0.2 N NaOH, briefly vortex and then centrifuge at 280 × *g* for 1 min. Incubate for 5 min at room temperature, then

add 990 μL prechilled HT1 buffer. This is a 10pM denatured library pool; dilute to 8pM by combining 800 μL 10pM denatured library pool with 190 μL prechilled HT1 buffer, and 10 μL 12.5pM denatured PhiX. Mix well.

6. Add 600–1000 μL of the denatured and diluted library pool + PhiX mix to the designated reservoir (labeled as LOAD SAMPLES) on the thawed reagent cartridge.
7. From the MiSeq Welcome Screen select "Sequence" and then follow the run set up steps.
8. Remove flow cell from storage buffer using plastic forceps. Rinse the flow cell under running distilled water to remove excess salts, dry thoroughly using a lint-free tissue, clean the flow cell glass with an alcohol wipe to ensure the glass is free for streaks, fingerprints, or fibers. Avoid touching the flow cell port gasket (black rubber area at the base of the flow cell). Allow to dry and load onto MiSeq following the screen prompts.
9. Load fresh PR2 bottle, and ensure the waste bottle is empty, following the screen prompts.
10. Load reagent cartridge into the chiller, following the screen prompts.
11. Change the sample sheet name to match bar code on reagent cartridge, review the run parameters and prerun check results.
12. Select Start Run (*see* **Note 29**).
13. When run is complete perform post run washes, as per standard MiSeq maintenance.

3.10 Data Analysis

Genome mapping using CLC Genomics Workbench from Qiagen (*see* **Note 30**).

1. Import data. For each sample, transfer the read_1 and read_2 FASTQ files from the MiSeq to your local computer (keep the file names). Import the files in CLC Genomics Workbench (keep the fastq format)—File > Import > Illumina.

   Select the following settings:

   - Select files (.fastq).
   - General options.
   - Paired reads.
   - Discard read names.
   - Illumina options.
   - Remove failed reads.
   - Paired reads information.
   - Paired-end (forward-reverse) minimum distance 1 and maximum distance 800.
   - Quality scores.

- Select: NCBI/Sanger or Illumina pipeline 1.8 and later.
- When imported the read_1 and read_2 files for each sample will be merged into one single file with (paired) added to the file name.

2. Quality Control (QC). Trim all read-pairs for the presence of ambiguous nucleotides and base call quality—Toolbox > NGS core tools > Trim Sequences.

   Select the following settings:

   - Select file (paired).
   - Trim using quality scores.
   - Limit 0.001 (= Phred score of 30).
   - Trim ambiguous nucleotides.
   - Maximum numbers of ambiguities 2.
   - (Adapter trimming was performed on the MiSeq so no further adapter trimming is necessary.)
   - Filter on length.
   - Maximum number of nucleotides in reads 1000.
   - Minimum number of nucleotides in reads 15.
   - Create report for inspection.

3. Reference mapping. Map all trimmed reads to a *C. trachomatis* reference genome:

   Identify the best matching (identical biovar/genotype) complete reference genome from GenBank NCBI nucleotide database (*see* **Note 31**). Select the FASTA format and download the full GenBank file to your local computer. Import the reference genome as a GenBank file (.gb)—File > Standard Import > Automatic import.

   Map each sample file containing the trimmed sequence reads to the selected reference(s)—Toolbox > NGS core tools > Map Reads to Reference.

   Select the following settings:

   - Select imported reference.
   - No masking.
   - Read alignment.

     Mismatch cost 2.

     Linear gap cost.

     - Insertion cost 3.
     - Deletion cost 3.

     Length fraction 0.5.

     Similarity fraction 0.8.

     Auto-detect paired distances.

- Non-specific map handling.
  - Map randomly.
- Create read track and report.

The mean read depth and fraction of the reference genome covered can be identified from the mapping report (*see* **Note 32**).

4. Consensus sequence. From each sample extract the consensus (majority) sequence based on the mapping file containing the trimmed sequence reads mapped to a reference.

   Select the following setting:

   - Select the mapped read track.
   - Threshold 5.
   - Insert "N" ambiguity symbols.
   - Conflict resolution "Vote".

For each sample the end result will be a complete (or close to complete) *C. trachomatis* genome generated based on reference mapping and majority base calling. Examples of how such genomes can be used for studies on population structure, in silico genotyping, identification of resistance mutation and variant detection can be found in Christiansen et al. [18].

4 Notes

1. After extraction *C. trachomatis* DNA can be quantified by qPCR, targeting the *C. trachomatis* plasmid and the genomic *omcB* gene, and human RNase-P as an endogenous control.
2. This protocol describes the use of the Agilent SureSelectXT 3 μg input methodology, in the majority of cases less than 3 μg of C. trachomatis DNA is obtained following extraction from clinical samples, we therefore use human DNA as a carrier to bulk the sample to 3 μg; however users may prefer to use the 200 ng method, which uses a smaller shearing volume (50 μL), and dilution of adapters 1:10, see http://www.chem.agilent.com/library/usermanuals/Public/G7530-90000.pdf for full details.
3. 70% ethanol should be made fresh daily using 100% molecular grade ethanol and ultrapure, molecular biology grade water.
4. This protocol describes the use of the Agilent SureSelect^{XT} 3 μg methodology, however since we first established this protocol other kits have come to market, including the SureSeq NGS Library Preparation Kit (500070, Oxford Gene Technology) which requires less bead clean-up steps.
5. 120-mer RNA baits spanning the length of the positive strand of 74 GenBank *C. trachomatis* reference genomes were

designed using an in-house PERL script developed by the PATHSEEK consortium. The specificity of the baits was verified by BLASTn searches against the Human Genomic Transcript database. The custom-designed *C. trachomatis* bait library was uploaded to SureDesign and synthesized by Agilent Technologies.

6. 10 M NaOH should be diluted to 0.2 N fresh, just before use.
7. Flow through contains Buffer AL or Buffer AW1 and is not compatible with bleach.
8. Add fresh deionized water to the Covaris tank and turn on chiller, allow to reach 5 °C before using.
9. Remove AMPureXP beads from fridge at least 30 min before use. Just before use shake beads vigorously to ensure homogenous and consistent color.
10. Bead pellet is completely dry when fine cracks can be observed. Do not over dry as this will decrease yield.
11. Remove TapeStation reagents from fridge 30 min prior to using.
12. If sample number is greater than 15 a 96-well plate can be used and multiple tapes.
13. Using a lower number of cycles will reduce the amount of duplication in the final sequencing library. The number of cycles required to achieve enough product for hybridization should be determined empirically.
14. 500–750 ng DNA from the Pre-Cap PCR is required for each hybridization, if there is not sufficient product from one PCR, repeat with another 15 μL aliquot of the Adapter Ligated sample, and after cleaning pool products.
15. If more than 8 samples are being hybridized use multiple plates and thermocycler blocks.
16. This step should be performed as quickly as possible to prevent excessive evaporation.
17. Use new strip caps to seal column 3; the structural integrity of the caps can be compromised during the previous incubation steps.
18. Excessive evaporation, such as the sample volume less than 20 μL following hybridization, can indicate suboptimal capture performance
19. Visually ensure that sample is mixing on the Nutator platform.
20. Pre-warm the required volume of SureSelect Wash Buffer #2 at 65 °C for at least 1 h before use.
21. Choose index combinations as per Illumina guidelines to ensure that indexes are balanced.
22. Using a lower number of cycles will reduce the amount of duplication in the final sequencing library. The number of

cycles required to achieve enough product for sequencing should be determined empirically.

23. We have found it to be more accurate to use the TapeStation data to establish the size of the DNA fragments, and the Qubit to measure the DNA concentration in order to calculate the molarity.
24. Calculate molarity of each Post-Index Amplified sample as:

$$\mu g DNA \times \frac{pmol}{660 pg} \times \frac{10^6 pg}{1 \mu g} \times \frac{1}{\mathrm{N}} = pmDNA$$

Where N is the average fragment size, and $\frac{pmol}{660 pg}$ is the average molecular weight of a nucleotide pair.

25. If creating a sample sheet in Excel *see* https://support.illumina.com/content/dam/illumina-support/documents/documentation/system_documentation/miseq/miseq-sample-sheet-quick-ref-guide-15028392-j.pdf for the parameters required.
26. PhiX control is helpful when sequencing low diversity samples; however, the choice to add it as a spike-in is down to the individual user. We found that it improved our cluster formation.
27. If you are using v3 chemistry then further dilution to 12.5pM is not required
28. Denatured PhiX libraries can be stored for up to 3 weeks at −20 °C
29. A 300 cycle run takes ~24 h to complete, monitor run for cluster number and performance.
30. http://www.clcbio.com/products/clc-genomics-workbench-direct-download/.
31. http://www.ncbi.nlm.nih.gov/nucleotide/.
32. When defining the *C. trachomatis* mean read depth obtained from vaginal swab samples, mask the duplicated rRNA regions as these regions have been found to have a significantly higher read depth compared to the rest of the *C. trachomatis* genome and will skew the results.

Acknowledgments

The PATHSEEK consortium received funding from the European Union's Seventh Framework Programme for research, technological development and demonstration under grant agreement No 304875. We acknowledge all the help from the other members of the consortium involved with the establishment of this methodology; particular thanks go to Helena Tutill, UCL.

References

1. Köser CU, Ellington MJ, Cartwright EJP et al (2012) Routine use of microbial whole genome sequencing in diagnostic and public health microbiology. PLoS Pathog 8:e1002824
2. Köser CU, Bryant JM, Becq J et al (2013) Whole-genome sequencing for rapid susceptibility testing of *M. tuberculosis*. N Engl J Med 369:290–292
3. Olsen RJ, Long SW, Musser JM (2012) Bacterial genomics in infectious disease and the clinical pathology laboratory. Arch Pathol Lab Med 136:1414–1422
4. WHO (2012) | Global incidence and prevalence of selected curable sexually transmitted infections - 2008. ISBN: 978 92 4 150383 9
5. WHO (2011) | Prevalence and incidence of selected sexually transmitted infections. *Chlamydia trachomatis*, *Neisseria gonorrhoeae*, syphilis and *Trichomonas vaginalis*. Methods and results used by WHO to generate 2005 estimates. ISBN: 978 92 4 150245 0
6. Mylonas I (2012) Female genital *Chlamydia trachomatis* infection: where are we heading? Arch Gynecol Obstet 285:1271–1285
7. Mariotti SP, Pascolini D, Rose-Nussbaumer J (2009) Trachoma: global magnitude of a preventable cause of blindness. Br J Ophthalmol 93:563–568
8. Blandford JM, Gift TL (2006) Productivity losses attributable to untreated chlamydial infection and associated pelvic inflammatory disease in reproductive-aged women. Sex Transm Dis 33:S117–S121
9. Burton MJ, Mabey DCW (2009) The global burden of trachoma: a review. PLoS Negl Trop Dis 3:e460
10. Pedersen LN, Herrmann B, Møller JK (2009) Typing *Chlamydia trachomatis*: from egg yolk to nanotechnology. FEMS Immunol Med Microbiol 55:120–130
11. Millman KL, Tavaré S, Dean D (2001) Recombination in the *ompA* gene but not the *omcB* gene of *Chlamydia* contributes to serovar-specific differences in tissue tropism, immune surveillance, and persistence of the organism. J Bacteriol 183:5997–6008
12. Psarrakos P, Papadogeorgakis E, Sachse K et al (2011) *Chlamydia trachomatis ompA* genotypes in male patients with urethritis in Greece: conservation of the serovar distribution and evidence for mixed infections with *Chlamydophila abortus*. Mol Cell Probes 25:168–173
13. Stothard DR, Boguslawski G, Jones RB (1998) Phylogenetic analysis of the *Chlamydia trachomatis* major outer membrane protein and examination of potential pathogenic determinants. Infect Immun 66:3618–3625
14. Harris SR, Clarke IN, Seth-Smith HMB et al (2012) Whole-genome analysis of diverse *Chlamydia trachomatis* strains identifies phylogenetic relationships masked by current clinical typing. Nat Genet 44:413–419. S1
15. O'Neill CE, Seth-Smith HMB, Van Der Pol B et al (2013) *Chlamydia trachomatis* clinical isolates identified as tetracycline resistant do not exhibit resistance in vitro: whole-genome sequencing reveals a mutation in *porB* but no evidence for tetracycline resistance genes. Microbiology 159:748–756
16. Seth-Smith HMB, Harris SR, Scott P et al (2013) Generating whole bacterial genome sequences of low-abundance species from complex samples with IMS-MDA. Nat Protoc 8:2404–2412
17. Seth-Smith HMB, Harris SR, Skilton RJ et al (2013) Whole-genome sequences of *Chlamydia trachomatis* directly from clinical samples without culture. Genome Res 23:855–866
18. Christiansen MT, Brown AC, Kundu S et al (2014) Whole-genome enrichment and sequencing of *Chlamydia trachomatis* directly from clinical samples. BMC Infect Dis 14:591
19. Depledge DP, Palser AL, Watson SJ et al (2011) Specific capture and whole-genome sequencing of viruses from clinical samples. PLoS One 6:e27805
20. Depledge DP, Kundu S, Jensen NJ et al (2014) Deep sequencing of viral genomes provides insight into the evolution and pathogenesis of varicella zoster virus and its vaccine in humans. Mol Biol Evol 31:397–409
21. Brown AC, Bryant JM, Einer-Jensen K et al (2015) Rapid whole genome sequencing of *M. tuberculosis* directly from clinical samples. J Clin Microbiol 53(7):2230–2237

Chapter 2

Characterization of Sinus Microbiota by 16S Sequencing from Swabs

Thad W. Vickery, Jennifer M. Kofonow, and Vijay R. Ramakrishnan

Abstract

New culture-independent microbiology methods are leading to a paradigm shift in our understanding of how the microbial community at the mucosal surface impacts sinonasal health and disease. Whereas traditional culture-based protocols were designed to identify specific pathogens in order to direct antibiotic therapies and eradicate bacteria, newer molecular techniques allow for the identification of both culturable and nonculturable bacteria in diverse communities. As a result of the recent explosion in the use of molecular techniques, we are gaining an understanding of how commensal bacteria may help modulate the host immune response and promote homeostasis. Here, we describe the general workflow of microbiome sequencing including the detailed methods for extracting mixed-community genomic DNA from sinonasal swabs, amplifying bacterial 16S rRNA genes using quantitative PCR, and preparing the samples for next-generation sequencing on the most commonly used sequencing platforms.

Key words Sinus swabs, DNA extraction, Bacterial 16S rRNA, Sinusitis, Culture-independent microbiology, Next generation sequencing, Microbiome

1 Introduction

The sinonasal cavity is known to harbor a rich and diverse variety of indigenous microbiota, which are increasingly appreciated for their role in promoting health and contributing to diseases such as acute and chronic rhinosinusitis [1–6]. Traditionally, the identification of microorganisms in the paranasal sinuses has relied on pure-culture techniques that often bias towards identifying more easily cultivated microorganisms and are limited in their capacity to identify particularly fastidious and/or anaerobic microbes [7]. Clinical microbiology has historically focused on identifying single pathogens that cause disease with the goal of targeting antibiotic therapy and eradicating bacteria. However, a new paradigm shift is underway emphasizing the role of commensal microbes within the community at the epithelial surface, and their ability to promote homeostasis and modulate the host mucosal immune response

Kimberly A. Bishop-Lilly (ed.), *Diagnostic Bacteriology: Methods and Protocols*, Methods in Molecular Biology, vol. 1616, DOI 10.1007/978-1-4939-7037-7_2,

[8–10]. Recent breakthroughs in culture-independent molecular methods have increased our ability to identify and characterize both culturable and nonculturable microbial communities including fastidious and anaerobic microbes [11–13]. These molecular techniques have been used to profile the microbiome within the healthy paranasal sinuses, which were found to possess a rich and diverse community of microbes including representatives of the phyla Firmicutes, Proteobacteria, Actinobacteria, and Bacteroides. Frequently identified species from the sinonasal niche include *Staphylococcus epidermidis*, *S. aureus*, *Propionibacterium* and *Corynebacterium* spp. among others [14].

Bacterial 16S ribosomal RNA (16S rRNA) is a component of the 30S ribosomal subunit in prokaryotic ribosomes. The genes encoding for the 16S rRNA contain highly conserved nucleotide sequences that allow for the design of universal PCR primers, which can be used to amplify the total 16S rRNA genes within a sample in an "unbiased" fashion. In addition to these conserved regions, 16S rRNA genes contain hypervariable regions that reflect species-specific signatures of particular bacterial taxa. Since the 16S gene is highly conserved among bacteria, it has become the standard for identification and phylogenetic classification [15–17]. Although 16S rRNA sequencing is often recognized as "unbiased," this is a subject of debate since the universal bacterial 16S DNA primers target only the bacterial domain and therefore do not identify other prokaryotic domains such as archaebacteria and unclassified prokaryotes [18]. Since the advent of next generation sequencing techniques, several databases are now available that catalog specific type strains of 16S rRNA genes. These can be compared at: http://www.ncbi.nlm.nih.gov/genomes/static/16S_comparison_help.html.

Here we will describe the general workflow and in-depth methods for characterizing the microbiome by sequencing the 16S rRNA gene from sinonasal swabs. Swabs provide an efficient and minimally invasive means to sample this mucosal surface microbiome. Collected sinus swabs may contain mucus, blood, and epithelial cells in addition to resident microbes. This mixed-community genomic DNA is isolated and used as the starting material for microbiome sequencing.

There are several methods reported in the literature to extract mixed-community DNA from various tissues [19]. Commercially available DNA isolation kits work well for isolating DNA from samples with low biomass, such as sinus swabs. In our experience, the amount of biomass present on the swab following sample collection dictates which DNA extraction method will return the cleanest sample for gene amplification and subsequent sequencing. We have found that samples with low biomass and very little blood contamination can be isolated using a commercial kit or directly precipitated and extracted using an abbreviated boiling method in nonionic detergent (NP-40), whereas samples with higher biomass including tissue and blood, are best extracted using a more rigorous phenol–chloroform solvent and grinding method [20–22].

Once DNA is isolated from other cellular components, the total bacterial 16S DNA in the sample is quantified by amplifying small-subunit (SSU) rRNA genes using fluorogenic universal primers and quantitative PCR (qPCR) to determine if there is bacterial DNA present in the sample and the optimal number of cycles for barcoding.

With sufficient quantities of 16S DNA confirmed, the individual samples are amplified again, this time using unique barcoding primers, which are primers containing short olignucleotide sequences specific to one individual sample. These barcodes are also designed to contain a short nucleotide sequence, called an adapter, which is specific to the sequencing platform. Each of the major sequencing platforms—including those offered by Illumina, Nanion Technologies, Oxford Nanopore Technologies, and Pacific Biosystems—utilizes unique oligonucleotide adapter sequences for recognizing samples that must be incorporated when designing barcode primers. Several open source software tools are available for designing barcoding primers [23, 24].

After each sample is labeled with a unique molecular barcode, the samples are pooled, normalized for concentration, and then sequenced. Particular rDNA sequences identified within a swab sample that correspond to a clone in the rRNA gene reference library are considered to exist within the community sampled within the sinonasal mucosal niche. Relative abundance of various bacterial taxa within a sample can be deduced using species-specific DNA or RNA-hybridization probes.

16S microbiome sequencing is a rapidly evolving field with new technology emerging that promises to increase the number and length of sequences that are achievable in a given sequencing run. As a result, there are several platform specific nuances and methods for analyzing sequence data that are beyond the scope of this methods paper. Here we aim to provide a succinct and detailed description of isolating mixed-community genomic DNA from sinonasal swabs and preparing the DNA for microbiome sequencing. Significant expertise is required to continue beyond this point, and the reader is instructed to other references in terms of selecting the appropriate regions to sequence [25, 26], appropriate assessment of sequence quality [27], database alignment and statistical methods [28–31].

2 Materials

In order to reduce the risk of ribonuclease or bacterial DNA/RNA contamination, all buffers and aliquot reagents should be prepared within a HEPA filtered laminar flow PCR hood. Ensure that all surfaces are wiped clean with 70% ethanol and disinfect the hood with 15 min of ultraviolet light before beginning. Whenever possible, use molecular grade reagents. Several of the listed reagents can be purchased premixed to limit potential for contamination.

2.1 Sample Collection

1. Single use, dry, sterile swabs. Several different swabs may be utilized, depending on the technique and desire. Flocked swabs can be utilized to obtain a more aggressive surface swab and incorporate host epithelial cells, if desired. We have utilized BD CultureSwab™ (Cat. No 220115) swabs to sample both awake and anesthetized patients, with sufficient tolerability and biomass yield to be able to sample subjects in a longitudinal fashion.

2.2 DNA Extraction from Low Biomass Swabs

1. Sterile 2 mL screw cap microcentrifuge tubes.
2. TEN Buffer: 10 mM Tris–HCl [pH 8.0], 1 mM EDTA, 1% NP40 nonionic detergent. Each sample requires 500 μL. For 100 mL, add 0.16 g Tris–HCL, 29.2 mg EDTA, and 1 mL NP40 to 99 mL of sterile molecular grade water.

2.3 DNA Extraction from High Biomass Swabs

1. Sterile 2 mL screw cap microcentrifuge tubes.
2. 2× Buffer B: 142.9 mM NaCl, 142.9 mM Tris–HCl [pH 8.0], 14.3 mM EDTA, 5.7% SDS. Each sample requires 700 μL. For 100 mL, combine 2.9 mL of 5 M NaCl stock, 14.3 mL 100× Tris–HCl stock, 418 mg EDTA, and 28.5 mL 20% SDS solution with 54.3 mL sterile molecular grade water.
3. Phenol–chloroform. Each sample requires 500 μL. For 100 mL, mix a 50 mL phenol and 50 mL chloroform.
4. Zirconia/Silica beads (0.1 mm): (BioSpec Cat. No 11079101z).
5. 7.5 M Ammonium acetate: Each sample requires 500 μL. For 100 mL, add 57.8 g ammonium acetate to 100 mL sterile molecular grade water.
6. Isopropanol (100%).
7. Glycogen (20 mg/mL): Each sample requires 2 μL. For 1 mL add 20 mg glycogen to 1 mL sterile molecular grade water.
8. Ethanol (70%): Each sample requires 250 μL. For 100 μL add 70 mL molecular grade ethanol to 30 mL sterile molecular grade water.
9. 1× Tris EDTA buffer (TE buffer): 10 mM Tris–HCl [pH 8.0], 1 mM EDTA. Each sample requires 50 μL. For 100 μL, combine 1 mL of 1 M Tris–HCl and 200 μL 0.5 M EDTA. Bring total volume to 100 mL using sterile molecular grade water.
10. Bench top mini-bead beater homogenizer (e.g., MagNA Lyser, Roche).

2.4 16S rDNA qPCR

1. PCR Mastermix: Several mixes are commercially available such as TaqMan® Universal PCR Master Mix, Applied Biosystems (Cat. Number: 4304437). Ensure that mix contains PCR buffer reagents (500 mM KCl, 100 mM Tris–HCl [pH 8.0], dNTP mix (2.5 mM each dNTP), 50 mM Magnesium Chloride, and Taq Polymerase.
2. SSU fluorogenic broad-specificity oligonucleotide primers for total bacterial 16S DNA (*see* **Note 1**).
3. 96-well PCR plate and 96-well plate cold block.
4. Plate sealing film for qPCR: e.g., ThermaSeal RT2RR™, Excel scientific (Cat. Number TS-RT2RR-100).
5. Standard 16S bacterial DNA template (*see* **Note 2**).
6. Mixed community genomic DNA (*see* **step 3**).
7. Thermocycler with Real-Time PCR capability.

2.5 Molecular Barcoding

1. Barcoding primers (*see* **Note 3**).
2. PCR Mastermix (Nova*Taq*™), 96-well PCR plate and 96-well plate cold block as described in Subheading 2.3, **item 1**.
3. Mixed community genomic DNA (*see* **step 3**).
4. Plate sealing film for PCR.
5. Thermocyler.

2.6 Agarose Gel Electrophoresis

1. Gel electrophoresis separation station and power supply.
2. TAE buffer (50× stock): 2 M Tris-Acetate, 0.05 M EDTA [pH 7.6]. Add 242 g Tris-base to 750 mL deionized RNAase free water in a 1 L graduated cylinder. Add 100 mL of 0.5 M EDTA to the solution and 57.1 mL glacial acetic acid. Adjust pH to 7.6 (at 25 °C) and bring final volume to 1 L. Filter solution through 0.22 μm filter to remove particles. Dilute to a 1× working solution when needed. (20 mL of 50× stock in 980 mL deionized RNAse free water).
3. Agarose.
4. Ethidium bromide: add 0.5 μg ethidium bromide per 1 mL agarose gel solution.
5. DNA Ladder (1 Kb).
6. Glycerol bromophenol blue loading dye (6×).
7. Long wavelength ultraviolet transilluminator.

2.7 DNA Normalization Reagents

1. SequalPrep® normalization plate kit, 96-well, Applied Biosystems (Cat. Number: A10510-01).

2.8 Sample Concentration and DNA Purification

1. Vacuum concentrator, e.g., Savant SpeedVac™ DNA Concentrator (Fisher).
2. Spin-column based DNA purification kit: e.g., DNA Clean & Concentrator™-5 (Zymo Research) or QIAquick PCR Purification columns (Qiagen).

2.9 Fluorometric DNA Quantification

1. Fluorometric DNA quantification kit: e.g., Qubit® dsDNA high sensitivity assay kit, Fisher (Cat. Number: Q32851) (*see* **Note 4**).
2. Fluorometer: e.g., Qubit® 3 fluorometer (Fisher).

3 Methods

Several commercially available DNA isolation kits such as the UltraClean® Microbial DNA Isolation Kit (Cat No: 12224-50) offered by Mo Bio Laboratories Inc. allow for efficient DNA extraction from swabs. Here we present two alternative protocols that may be used in addition to commercially available kits. Spin-column based kits use a solid-phase silica gel membrane that binds DNA when the pH and salt concentrations are optimal. This is easily saturated when samples contain large amounts of genomic DNA. The following two methods do not use solid phase sorbents and therefore are scalable allowing for isolation of larger amounts of genomic DNA. Each method presented below has been shown to yield comparable results [20, 22]. We recommend solvent-based phenol–chloroform extraction with mechanical bead-beating for samples with large amounts of tissue or blood (*see* Table 1 for examples of concentrations and Table 2 for examples of purity measurements obtained). Ensure that appropriate negative and positive controls are included in the DNA isolation. Sterile swabs should be extracted alongside patient samples to serve as negative controls. A positive control can be created by swabbing an agar plate containing bacterial colonies.

In order to reduce nucleic acid contamination, conduct all DNA isolation steps and PCR amplification preparation within a HEPA filtered laminar flow hood. Ensure that all surfaces are wiped clean with 70% ethanol and disinfect the hood with 15 min of ultraviolet light before beginning.

3.1 Swab Collection and Storage

1. Sample sinonasal mucosa by rotating swab clockwise at least five times in the region of interest, until the swab is visually saturated. Transfer swab to original collection tube and store on ice (*see* **Note 5**).
2. Within 2 h of collection, transfer the swabs to a sterile 2 mL screw cap microcentrifuge tube in a disinfected laminar flow hood. The excess swab may be trimmed using isopropanol-treated scissors or carefully snapped off to allow the microcentrifuge tube lid to close.
3. Store swabs at −80 °C until further processing.

Table 1

An overnight culture of methicillin-resistant *Staphylococcus aureus* (MRSA) in tryptic soy broth was divided into identical 1 mL aliquots and subjected to various conditions for DNA extraction in duplicate: Negative (no bacteria), Vortex (vortex only), Boil (vortex and boil only), and bead beat at 3000 rpm (845 × *g*), 5000 rpm (2348 × *g*), or 7000 rpm (4602 × *g*) vs. 30, 60, or 99 s on a Roche MagNA Lyser bead beater[a]

DNA concentration (ng/μL)						
				Speed (rpm)		
Negative	Vortex	Boil		3000 (845 × *g*)	5000 (2348 × *g*)	7000 (4602 × *g*)
4.9	101.3	187.4	30 s	180.8	184.8	205
−22.8	107.3	180.8	30 s	175.6	183.6	210
			60 s	202.9	176.7	224.6
			60 s	188.6	194	213.1
			99 s	249.9	213.2	242.1
			99 s	240.9	234.5	231.4

After extraction each sample was centrifuged for 5 min at 10,000 rpm (9391 × *g*) and the supernatant containing DNA was saved. DNA concentrations (ng/μL) were measured using a NanoDrop ND-1000 spectophotometer

[a]Samples subjected to bead beating were also vortexed and boiled beforehand

Table 2

DNA purity (from the samples described in Table 1) was determined by measuring the ratio of absorbance at 260 nm and 280 nm using a NanoDrop ND-1000 spectrophotometer

Absorbance ratio 260/280 nm						
				Speed (rpm)		
Negative	Vortex	Boil		3000 (845 × *g*)	5000 (2348 × *g*)	7000 (4602 × *g*)
−0.23	1.42	1.71	30 s	1.83	1.82	1.77
0.46	1.43	1.75	30 s	1.86	1.84	1.71
			60 s	1.79	1.84	1.68
			60 s	1.85	1.79	1.79
			99 s	1.76	1.76	1.81
			99 s	1.76	1.76	1.71

A 260/280 ratio of ~1.8 is generally considered "pure" for DNA

3.2 DNA Isolation of Low Biomass Swabs

1. Add 500 μL of TEN buffer to the dried swab heads in the same 2 mL screw-cap tube used for storage.
2. Place tubes in 95 °C incubator bath for 10 min to lyse cells. After 10 min, remove each sample and vortex for 30 s. Incubate tubes at 95 °C for an additional 10 min.

3. Remove swabs using isopropanol-treated forceps and place samples on ice or store at −20 °C until PCR.
4. If additional extraction is desired, add 0.25 g of 0.1 mm zirconium beads and bead beat for 60 s at 6000 rpm (3381 × *g*) (*see* **Note 6**).
5. Centrifuge the samples at 10,000 rpm (9391 × *g*) for 5 min and transfer supernatant to a clean 2 mL screw-cap tube. Samples can be stored at −20 °C until use.

3.3 DNA Extraction of High Biomass Swabs

3.3.1 Cell Lysis

1. Add 700 μL of 2× buffer B and 0.25 g of 0.1 mm zirconium beads to each specimen tube (*see* **Notes 7** and **8**).
2. Mix by inverting several times.
3. Add 500 μL phenol–chloroform to each sample (*see* **Note 9**).
4. Place samples in a mini bead-beater homogenizer for 60 s at 6000 rpm (3381 × *g*). At the completion of the bead-beating cycle, the cotton tip of each swab and any visible tissue should be completely homogenized.
5. Centrifuge samples at 10,000 rpm (9391 × *g*) at room temperature for 5 min to deposit beads and sample debris into pellet.
6. Two phases (an upper aqueous phase and a lower organic phase and interface) will partition following centrifugation. Collect the upper aqueous phase using a pipette and transfer to a fresh 2 mL tube.
7. Repeat this process until the interface (thin area between the upper aqueous and lower organic phase) is clear. Collect only the upper aqueous phase and transfer to a fresh 2 mL tube.

3.3.2 Precipitation and DNA Isolation

8. Add 300 μL 7.5 M ammonium acetate, 500 μL 100% isopropanol, and 2 μL of 20 mg/mL glycogen (*see* **Note 10**).
9. Incubate samples at −80 °C for 2 h (*see* **Note 11**).
10. Centrifuge samples at 10,000 rpm (9391 × *g*) for 30 min at 4 °C.
11. Decant supernatant by drawing off liquid from each sample using a pipette without disrupting the pellet. Discard supernatant. Not all samples will have a visible pellet.
12. Add 250 μL of cold 70% ethanol to each sample. Vortex briefly and centrifuge for 1 min at 10,000 rpm (9391 × *g*). Carefully remove supernatant without disrupting pellet. Discard supernatant (*see* **Note 12**).
13. Evaporate solvent to dryness in a vacuum concentrator (e.g., SpeedVac™, Thermo Scientific) or if unavailable, samples may be evaporated in a UV-sterilized laminar flow hood.
14. Resuspend pellets in 50 μL of 1× TE buffer. Tightly close the microcentrifuge tubes prior to storage to limit evaporation. Concentrated DNA Samples can be stored at −80 °C.

3.4 Measuring Total Bacterial 16S DNA by qPCR

1. Prepare the mastermix in a 2 mL microcentrifuge tube on a cold block. The total PCR reaction volume is 20 μL per sample. Use the appropriate dilution of master mix stock according to the manufacturer. For example, aliquot 10 μL 2× master mix, 7 μL of sterile RNAase free water, and 1 μL of 20× forward and reverse primers. Prepare sufficient excess mastermix to allow for appropriate controls including negative controls (DNA extracts from sterile swabs and RNAase-free water alone), positive extraction controls (bacterial colonies swabbed from agar), and standard curve wells.
2. Pipette 18 μL of the mastermix into a corresponding well in a 96-well plate on a cold block.
3. Add 2 μL of extracted genomic DNA from each swab sample to its corresponding well on the 96-well plate.
4. Add appropriate amount of genomic or plasmid based 16S bacterial DNA template to create standard curve ranging from 10^8 to 10^1 copies (*see* **Note 2**).
5. Cover the plate tightly with qPCR compatible plate sealing film, vortex briefly, and centrifuge plate briefly to ensure sample is completely mixed.
6. Transfer 96-well plate to thermocycler and initiate qPCR with 7-min denaturation at 95 °C followed by 40 cycles of 15 s at 95 °C, and 1 min at 60 °C, including a plate-reading step at the end of each cycle.
7. After the qPCR run completes, use cycle threshold values (C_t) to determine the appropriate number of cycles necessary to amplify DNA in the subsequent barcoding step (*see* **Note 13**). *See* Table 3 for example C_t values obtained.

3.5 Molecular Barcoding

1. Aliquot 20× forward and reverse barcoding primers into 96-well plate with each well containing a unique barcoded sequence with the same 5′ sequencing platform adapter oligonucleotide and the 16S DNA universal primer sequence. Document the location of each barcode on the 96-well primer plate. Cover the primer plate with aluminum sealing foil and store at −20 °C when not in use.
2. Prepare the mastermix (Nova*Taq*™) for the barcoding reaction in a 2 mL microcentrifuge tube on a cold block. The total PCR reaction volume is 30 μL per sample. Use the appropriate dilution of master mix stock according to the manufacturer as performed in Subheading 3.4, **step 1**. Distribute the master mix (without primers) and sterile water into corresponding wells on a new 96-well plate on a cold block.
3. Using a multichannel pipette, transfer 1.5 μL of each forward and reverse barcoding primers from the 96-well primer plate to

Table 3

Cycle threshold (C_t) values for qRT-PCR of the total bacterial 16S gene for DNA samples (described in Table 1)

16S qRT-PCR C_t Values						
				Speed (rpm)		
Negative	Vortex	Boil		3000 (845 × *g*)	5000 (2348 × *g*)	7000 (4602 × *g*)
36.35	31.25	30.87	30 s	28.21	21.14	19.41
36.94	28.72	32.7	30 s	28.24	24.25	19.1
			60 s	21.57	17.15	15.39
			60 s	21.7	17.42	16.13
			99 s	17.28	16.39	15.17
			99 s	18.05	16.02	15.36

A lower C_t value correlates with a higher concentration of bacterial DNA. (Negative controls for the PCR reagents did not contain any detectable levels of bacterial DNA)

the corresponding wells on the sample plate. Prepare sufficient excess mastermix to allow for appropriate controls including negative controls (DNA extracts from sterile swabs and RNAse-free water alone) and positive extraction controls (bacterial colonies swabbed from agar).

4. Add 4 μL of mixed-community genomic DNA extract from each swab sample to its corresponding well on the 96-well sample plate.
5. Cover the 96-well sample plate with an adhesive plate seal, briefly vortex, and centrifuge to concentrate sample in the bottom of each well.
6. Transfer 96-well plate to thermocycler and initiate PCR with 7-min denaturation at 95 °C followed by 30 s at 94 °C, 30 s at 54.3 °C, and 60 s at 72 °C for the appropriate number of cycles. Following the last cycle, initiate a final extension of 10 min at 72 °C. Determine the optimal number of cycles based on previous C_t results (*see* **Note 13**). Typical cycles used for barcoding sinus samples range between 26 and 30.
7. After the thermocycling sequence is complete, plate may be stored at −20 °C.

3.6 Confirmation of Amplification with Gel Electrophoresis

8. Since florescent probes would interfere with the sequencing process, the best way to ensure sample 16S gene amplification is with an agarose DNA gel. Prepare a 2% agarose gel by mixing agarose with the appropriate volume of 1× TAE buffer for the gel cassette you will be using. For example, mix 2 g agarose in 100 mL 1× TAE buffer.

9. Microwave the mixture for 1–3 min until the agarose is completely dissolved and the solution is gently boiling. Carefully remove the solution from the microwave and let cool on the bench top for 5 min.
10. Add ethidium bromide to the mixture to a final concentration of approximately 0.2–0.5 μg/mL.
11. Pour the gel into the gel tray and position the well comb in place. Allow gel to set at 4 °C or room temperature until completely solidified.
12. Place gel into gel box and completely submerge in 1× TAE buffer. Remove well comb (*see* **Note 14**).
13. Add 1 Kb DNA molecular weight ladder to the first well on agarose gel.
14. Mix 10 μL of loading buffer with 5 μL of barcoded DNA and transfer to well on agarose gel. Repeat for all barcoded samples. Note the position of each sample on the gel relative to its corresponding location on the 96-well plate.
15. Run the gel at 100 V until the dye migrates 75–80% through the gel. Turn off power supply and remove gel.
16. Visualize DNA bands using long-wavelength ultraviolet light (*see* **Note 15**).

3.7 DNA Concentration Normalization and Sample Pooling

1. Load the remaining 25 μL of amplicon from each sample in the barcoding reaction plate into a corresponding well on a SequalPrep™ 96-well normalization plate (*see* **Note 16**).
2. Follow protocol for DNA binding, washing, and elution steps as noted in the SequalPrep® normalization plate kit protocol such that each sample is suspended in 20 μL of elution buffer.
3. Pool all samples into a clean 2 mL microcentrifuge tube and transfer to vacuum concentrator. Evaporate until ~200 μL elution buffer remains. Large sample sets may require more than one microcentrifuge tube to concentrate DNA. Concentrated pooled amplicons may be stored at −80 °C.

3.8 Purify and Measure DNA Concentration

1. Transfer 100 μL pooled and concentrated amplicons to a clean 2 mL microcentrifuge tube. Follow Clean & Concentrator™-5 kit (or similar kit) instructions. Elute sample in 30 μL of DNA elution buffer (*see* **Note 17**).
2. Follow fluorometric DNA assay kit protocol to determine concentration of purified, pooled and barcoded amplicon sample and dilute DNA to the appropriate concentration for the specific sequencing platform being used.

4 Notes

1. Primers for total bacterial DNA qPCR are generated using amplicons of the 16S rRNA gene with fluorogenic reporters (such as HEX or FAM) as described by Nadkarni et al. [32]. There are 9 hypervariable regions contained within the bacterial 16S rRNA gene. Primers that target specific segments of the variable region may preferentially target certain taxa with variability contained in the selected region [25]. There are several examples in the literature of 16S rRNA amplification from upper respiratory swabs using primers that target the variable regions V1-V3 for 16S rRNA genes 8F, 805R, 515F, and 1391R among others [20, 22, 33].
2. In order to perform qPCR on mixed-community DNA extracted from swab samples, a standard curve must be generated with known amounts of bacterial gene template. Several approaches have been reported using genomic DNA or DNA derived from plasmids to generate bacterial DNA templates for qPCR. We utilize a plasmid vector to produce Lachnospiraceae 16S gene template and construct standard curves ranging from 10^8 to 10^1 copies by making tenfold serial dilutions of the template stock.
3. Barcoding primers are similar to the universal primers designed to amplify the 16S rRNA gene. However, barcoding primers contain short oligonucleotide sequences that are unique to an individual sample that are later used to identify the sample after all samples are pooled for sequencing. Additionally, barcoding primers are designed to contain platform-specific 5′ adapter oligonucleotide sequences that are specific to the sequencing platform being used. Each sequencing platform provides detailed amplicon sequencing protocols that provide the adaptor sequences. There are several software programs that facilitate the efficient design of barcoding primers. For example, BARCRAWL and BARTAB are open source software tools available for designing barcoding primers [23, 24].
4. Many of the sequencing platforms currently in use are exquisitely sensitive to DNA overloading. Therefore, it is important that a reliable and accurate method is used to quantify the DNA prior to loading into the sequencer. Traditional methods using absorbance ratios at 260 nm and 280 nm (A_{260}/A_{280}) are not sensitive at very low concentrations of DNA, and absorbance varies with pH. Using a fluorometric method for DNA quantification allows for a more specific and sensitive measurement. There are several commercially available fluorometric kits that utilize Hoechst 33258 or PicoGreen.
5. Since the bacterial niche is variable from different sites within the nose and paranasal sinuses, it is critical to sample the spe-

cific region of interest using a consistent and standardized technique. In clinical settings where multiple clinicians are collecting swabs, standard collection protocols should be used. Sinus samples that come in contact with the mucosa of the anterior nares upon removal should be discarded.

6. When using a bead beater to extract DNA, the conditions should first be optimized in order to determine the most appropriate length and speed that extracts the most DNA without damaging it. An easy way to do this is to prepare several identical 1 mL aliquots of the same overnight culture in liquid media, pellet the cells by centrifugation, and resuspend in 1× TEN with 0.1 mm zirconia beads. Boil as previously instructed and then bead beat at different combinations and variations of time and speed in replicates. Pellet the debris by centrifugation and save the supernatant containing genomic DNA. Optimal conditions are determined by measuring the optical density at 260 nm for concentration, recording the OD_{260}/OD_{280} ratio for purity, and deciphering which condition yields the best Ct values.
7. In order to ensure lysis of all bacterial cells, some protocols add lysozyme to the buffer B (final concentration 5 mg/mL) and incubate for 30 min at 37 °C to help disrupt gram-positive cell walls. This is followed by the addition of SDS (final concentration 0.3%) and proteinase K (final concentration 2 mg/mL) and further incubation at 50 °C for 30 min prior to bead homogenization.
8. Zirconia/silica beads are spherical and do not fracture under extremely high shearing forces. These will effectively shear cell walls but leave the DNA intact.
9. Phenol–chloroform is a mixture of organic solvents that trap non-nucleic acid components of lysed cells such as lipids, proteins, and polysaccharides.
10. This step directly leads to the precipitation of DNA from the solution. DNA is insoluble in isopropanol. Isopropanol precipitates DNA at much lower concentrations compared to other alcohols such as ethanol. Ammonium acetate is added to increase the monovalent cation salt concentration and aids in precipitation of DNA and decreases precipitation of free nucleotides. Glycogen is a large polysaccharide that is added as an inert carrier (co-precipitant), which helps trap nucleic acids by bulk to increase DNA recovery during centrifugation—especially in swab samples containing very low concentrations of DNA.
11. Precipitation of DNA at lower temperatures slows the precipitation by increasing the dielectric constant and the viscosity of the mixture. Decreasing the temperature helps DNA retain a native shape and limits the actions of endogenous nucleases

during the precipitation. However, the tradeoff of performing precipitations at lower temperatures is lower DNA yields and some co-precipitation of salt components.

12. Washing the pellet with 70% ethanol helps remove any salts that co-precipitated with the genomic DNA.
13. The cycle threshold (C_t) is the number of cycles necessary for the fluorescence in the qPCR sample to exceed the background level. Lower C_t values reflect higher levels of 16S amplification within a sample. Sinus swabs often yield abundant 16S DNA with typical C_t values ranging from 12 to 30. After the completion of the qPCR run, samples with lower C_t values must be diluted to normalize the concentration across samples prior to barcoding. For example, if the mean C_t value for a group of sinus swab samples is 25, samples with C_t values of 12 must be diluted 1:2 in order to more closely match the amount of 16S DNA contained in the remaining samples. Samples with C_t values of 34 or higher typically do not contain amplified DNA beyond background levels. Samples with C_t values greater than the mean should proceed to barcoding without concentrating. The minimum number of cycles should be used to maximize the variability of amplified DNA. The qPCR run can also be used to determine the optimal number of cycles for subsequent barcoding, which for sinus swab samples is typically 26–30 cycles.
14. Ethidium bromide (EtBr) is a DNA intercalating agent and known mutagen. Use appropriate PPE when working with EtBr and dispose of waste properly. Adding EtBr allows for the DNA bands to be visualized using ultraviolet light. EtBr is also positively charged and therefore tends to run in the opposite direction of DNA. In order to minimize the separation of EtBr from DNA bands, add a small amount of EtBr to the running buffer (approximately 0.1 μg/mL).
15. Normally two bands are visualized on the agarose gel including a single band corresponding to the length of the target amplicon (length of forward primer minus length of reverse primer). The lowest molecular weight band (furthest migrated through the gel) is composed of primer dimers and loading buffer components. Occasionally, there is a higher molecular weight band that separates above the desired sequence. This reflects nonspecific binding of the primers to off-target genes. If the nonspecific binding band is substantial, an additional gel purification step may be added wherein the band containing the desired DNA is carefully cut out from the gel and purified. Samples that do not show an amplified band corresponding to 16S DNA should correspond to DNA extracts with high C_t values in the total bacteria qPCR run. Samples with low C_t values by qPCR that subsequently do not amplify in the barcoding PCR reaction should be troubleshooted and repeated.

16. Several methods for DNA normalization are available including direct quantification (using NanoDrop technology), size restricted DNA quantification (such as the QIAxcel from Qiagen), and quantitative DNA binding (such as SequalPrep kit from Invitrogen). The SequalPrep kit has been shown to produce the best normalized barcoded amplicon pool and offers a more efficient workflow than other alternatives [34].
17. The barcoding PCR reaction products contain primer dimers, salts, enzymes, and other impurities. Although the SequalPrep normalization step removes several of these impurities, there may be primer dimers and cationic salts that remain bound to the DNA. Therefore, we implement a quick spin-column based purification step prior to sequencing.

Acknowledgments

The authors would like to acknowledge Diana Ir for generously sharing her knowledge of 16S microbiome characterization, reviewing early drafts and providing valuable feedback that improved the manuscript significantly. In addition, we would like to thank Dr. Daniel N. Frank, Ph.D., for his generous support and collaboration. Research reported in this publication was supported by the National Institute On Deafness And Other Communication Disorders of the National Institutes of Health under award numbers K23DC014747 (V.R. Ramakrishnan) and T32DC01228003 (T.W. Vickery), as well as the Flight Attendants Medical Research Institute grant CIA13006 (V.R. Ramakrishnan and D.N. Frank). The content is solely the responsibility of the authors and does not necessarily represent the official views of the National Institutes of Health or the Flight Attendants Medical Research Institute.

References

1. Ramakrishnan VR, Hauser LJ, Frank DN (2016) The sinonasal bacterial microbiome in health and disease. Curr Opin Otolaryngol Head Neck Surg 24:20–25
2. Hamilos DL (2014) Host-microbial interactions in patients with chronic rhinosinusitis. J Allergy Clin Immunol 133:640–653. e4
3. Feazel L, Robertson C, Ramakrishnan VR, Frank DN (2012) Staphylococcus aureus and microbiome diversity in chronic rhinosinusitis. Laryngoscope 122(2):467–472
4. Ramakrishnan VR, Hauser LJ, Feazel LM, Ir D, Robertson CE, Frank DN (2015) Sinus microbiota varies among chronic rhinosinusitis phenotypes and predicts surgical outcome. J Allergy Clin Immunol 136(2):334–342
5. Vickery TV, Ramakrishnan VR (2017) Bacterial pathogens and the microbiome. Otolaryngol Clin N Am 50(1):29–47
6. Lee JT, Frank DN, Ramakrishnan VR (2016) Sinus microbiome. Am J Rhinol Allergy 30(1):3–16
7. Hauser LJ, Feazel LM, Ir D et al (2015) Sinus culture poorly predicts resident microbiota. Int Forum Allergy Rhinol 5:3–9
8. Kamada N, Seo S-U, Chen GY et al (2013) Role of the gut microbiota in immunity and inflammatory disease. Nat Rev Immunol 13:321–335
9. Tabas I, Glass CK (2013) Anti-inflammatory therapy in chronic disease: challenges and opportunities. Science 339:166–172

10. McLoughlin RM, Mills KHG (2011) Influence of gastrointestinal commensal bacteria on the immune responses that mediate allergy and asthma. J Allergy Clin Immunol 127: 1097–1108
11. Feazel LM, Robertson CE, Ramakrishnan VR et al (2012) Microbiome complexity and *Staphylococcus aureus* in chronic rhinosinusitis. Laryngoscope 122:467–472
12. Abreu NA, Nagalingam NA, Song Y et al (2012) Sinus microbiome diversity depletion and *Corynebacterium tuberculostearicum* enrichment mediates rhinosinusitis. Sci Transl Med 4:151ra124
13. Stephenson MF, Mfuna L, Dowd SE et al (2010) Molecular characterization of the polymicrobial flora in chronic rhinosinusitis. J Otolaryngol Head Neck Surg 39:182–187
14. Ramakrishnan VR, Feazel LM, Gitomer SA et al (2013) The microbiome of the middle meatus in healthy adults. PLoS One 8:e85507
15. Ward DM, Weller R, Bateson MM (1990) 16S rRNA sequences reveal numerous uncultured microorganisms in a natural community. Nature 345:63–65
16. Weisburg WG, Barns SM, Pelletier DA et al (1991) 16S ribosomal DNA amplification for phylogenetic study. J Bacteriol 173:697–703
17. Turnbaugh PJ, Ley RE, Hamady M et al (2007) The human microbiome project. Nature 449:804–810
18. Woese CR, Fox GE (1977) Phylogenetic structure of the prokaryotic domain: The primary kingdoms. Proc Natl Acad Sci U S A 74:5088–5090
19. Yuan S, Cohen DB, Ravel J et al (2012) Evaluation of methods for the extraction and purification of DNA from the human microbiome. PLoS One 7:e33865
20. Frank DN, Feazel LM, Bessesen MT et al (2010) The human nasal microbiota and *Staphylococcus aureus* carriage. PLoS One 5:e10598
21. Frank DN, Wysocki A, Specht-Glick DD et al (2009) Microbial diversity in chronic open wounds. Wound Repair Regen 17:163–172
22. Frank DN, Spiegelman GB, Davis W et al (2003) Culture-independent molecular analysis of microbial constituents of the healthy human outer ear. J Clin Microbiol 41:295–303
23. Hamady M, Walker JJ, Harris JK et al (2008) Error-correcting barcoded primers for pyrosequencing hundreds of samples in multiplex. Nat Methods 5:235–237
24. Frank DN (2009) BARCRAWL and BARTAB: software tools for the design and implementation of barcoded primers for highly multiplexed DNA sequencing. BMC Bioinformatics 10:362
25. Chakravorty S, Helb D, Burday M et al (2007) A detailed analysis of 16S ribosomal RNA gene segments for the diagnosis of pathogenic bacteria. J Microbiol Methods 69:330–339
26. Rajendhran J, Gunasekaran P (2011) Microbial phylogeny and diversity: Small subunit ribosomal RNA sequence analysis and beyond. Microbiol Res 166:99–110
27. Zhou Q, Su X, Ning K (2014) Assessment of quality control approaches for metagenomic data analysis. Sci Rep 4:6957
28. Ewing B, Hillier L, Wendl MC et al (1998) Base-calling of automated sequencer traces using phred. I Accuracy assessment. Genome Res 8:175–185
29. Edgar RC, Haas BJ, Clemente JC et al (2011) UCHIME improves sensitivity and speed of chimera detection. Bioinformatics 27: 2194–2200
30. Schloss PD, Westcott SL (2011) Assessing and improving methods used in operational taxonomic unit-based approaches for 16S rRNA gene sequence analysis. Appl Environ Microbiol 77:3219–3226
31. Quast C, Pruesse E, Yilmaz P et al (2012) The SILVA ribosomal RNA gene database project: improved data processing and web-based tools. Nucleic Acids Res 41:D590–D596
32. Nadkarni MA, Martin FE, Jacques NA et al (2002) Determination of bacterial load by real-time PCR using a broad-range (universal) probe and primers set. Microbiology 148: 257–266
33. Eden PA, Schmidt TM, Blakemore RP et al (1991) Phylogenetic analysis of *Aquaspirillum magnetotacticum* using polymerase chain reaction-amplified 16S rRNA-specific DNA. Int J Syst Bacteriol 41:324–325
34. Harris JK, Sahl JW, Castoe TA et al (2010) Comparison of normalization methods for construction of large, multiplex amplicon pools for next-generation sequencing. Appl Environ Microbiol 76:3863–3868

Chapter 3

Molecular Subtyping of *Salmonella* Typhimurium with Multiplex Oligonucleotide Ligation-PCR (MOL-PCR)

Véronique Wuyts, Wesley Mattheus, Nancy H.C. Roosens, Kathleen Marchal, Sophie Bertrand, and Sigrid C.J. De Keersmaecker

Abstract

A multiplex oligonucleotide ligation-PCR (MOL-PCR) assay is a valuable high-throughput technique for the detection of bacteria and viruses, for characterization of pathogens and for diagnosis of genetic diseases, as it allows one to combine different types of molecular markers in a high-throughput multiplex assay. A MOL-PCR assay starts with a multiplex oligonucleotide ligation reaction for detection of the molecular marker, followed by a singleplex PCR for signal amplification and analysis of the MOL-PCR products on a Luminex platform. This last step occurs through a liquid bead suspension array in which the MOL-PCR products are hybridized to MagPlex-TAG beads.

In this chapter, we describe the complete procedure for a MOL-PCR assay for subtyping of *Salmonella enterica* subsp. *enterica* serovar Typhimurium (*S.* Typhimurium) and its monophasic variant *S.* 1,4[5],12:i:- from DNA isolation through heat lysis up to data interpretation through a Gödel Prime Product. The subtyping assay consists of 50 discriminative molecular markers and two internal positive control markers divided over three MOL-PCR assays.

Key words MOL-PCR, *Salmonella* Typhimurium, Subtyping, Luminex, Bead suspension array, Gödel Prime Product, High-throughput assay

1 Introduction

Multiplex oligonucleotide ligation-PCR (MOL-PCR) [1–3] allows for examining different types of molecular markers, such as unique sequences, single nucleotide polymorphisms (SNPs), insertions, and deletions, in a single assay. The technique consists of a marker detection step through a multiplex oligonucleotide ligation reaction, a signal amplification step via singleplex PCR, and an analysis (readout) step on a Luminex device, which requires the hybridization of the MOL-PCR products to MagPlex-TAG beads and an incubation with streptavidin-R-phycoerythrin (SAPE) for binding of the necessary fluorescent reporter to the biotin label on the MOL-PCR products (*see* Fig. 1). Each marker

Kimberly A. Bishop-Lilly (ed.), *Diagnostic Bacteriology: Methods and Protocols*, Methods in Molecular Biology, vol. 1616, DOI 10.1007/978-1-4939-7037-7_3,

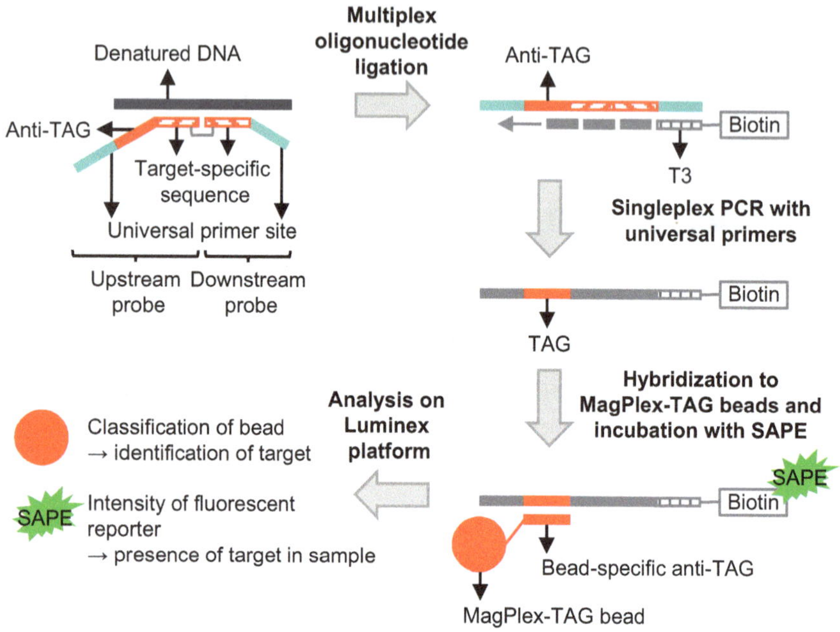

Fig. 1 Schematic overview of a multiplex oligonucleotide ligation-PCR (MOL-PCR) assay for one target. Detailed steps are described in the introduction. *SAPE* streptavidin-R-phycoerythrin

is detected through two probes which hybridize exactly adjacently on the target DNA. The upstream probe includes a 5′ T7 primer sequence, an internal TAG sequence (used to discriminate each marker) and a 3′ marker-specific sequence. For each marker, a different TAG sequence is used. The downstream probe includes a 5′ marker-specific sequence and a 3′ reverse complement of a T3 primer sequence. It is modified at 5′ with a phosphate which is required for the ligation of the upstream and downstream probe through a phosphodiester bond. Ligation of corresponding upstream probes and downstream probes results in various DNA strands which serve as templates in the subsequent singleplex PCR with universal primers T7 and T3. The T3 primer is biotinylated so that the MOL-PCR products can be fluorescently labeled with SAPE. The biotinylated amplicons generated with the T3 primer carry an anti-TAG (*see* Fig. 1), which is the complement of the TAG sequence in the upstream probe and which is also the complement of the TAG sequence that is coupled to MagPlex-TAG beads. MagPlex-TAG beads are colored in different shades of red and each set of MagPlex-TAG beads with a specific red color, which is called a bead region, is coupled with a specific TAG sequence. As each marker corresponds to a different TAG sequence, each marker also corresponds to a different bead region. The Luminex platform can thus identify the marker through recognition of the red color, or the bead region, of the MagPlex-TAG bead. The presence or absence of a hybridized MOL-PCR product, thus the presence or absence of the marker in the sample, is observed through the fluorescent signal of the SAPE bound to the biotin label of the MOL-PCR product.

We have designed a high-throughput MOL-PCR assay for subtyping of *Salmonella enterica* subsp. *enterica* serovar Typhimurium (*S.* Typhimurium) [4, 5], which is a frequently isolated foodborne pathogen [6, 7] and for which subtyping is thus of outmost significance for surveillance and outbreak investigations. This assay combines 52 molecular markers which are divided over three MOL-PCR assays, i.e., MOL-PCR_1, MOL-PCR_2, and MOL-PCR_SNP (*see* Table 1). The markers include prophage genes, amplified fragment length polymorphism (AFLP) elements, *Salmonella* genomic island 1 (SGI1), antibiotic resistance genes, single nucleotide polymorphisms (SNPs), multiple-locus variable-number of tandem repeats analysis (MLVA) locus STTR10, allantoinase gene *allB*, and phase 2 flagellar gene *fljB* for detection of the monophasic variant of *S.* Typhimurium, i.e., *S.* 1,4,[5],12:i:-. Two internal positive control markers check the DNA template and are specific for all *Salmonella* spp. (invasive gene *invA*) and for *S.* Typhimurium (SNP in the β subunit of bacterial RNA polymerase encoding gene *rpoB*).

The procedure starts with preparation of the DNA template through heat lysis of a single colony of a *S.* Typhimurium isolate. After the actual MOL-PCR assays, the multiplex results are summarized in a MOL-PCR profile through calculation of the Gödel Prime Product (GPP) [8, 9]. This MOL-PCR profile acts a barcode [10] for the isolate and can be used to easily compare the MOL-PCR profiles of different *S.* Typhimurium isolates.

Table 1
Type of markers and division over MOL-PCRs in the subtyping assay [4]

MOL-PCR	Internal PC	Prophage	AFLP	SGI1	Antibiotic resistance	SNP	Other	Plex
MOL-PCR_1	*invA* *rpoB*	8	2	2	4	–	STTR10 *allB*	20
MOL-PCR_2	*rpoB*	9	10	–	1	–	*fljB*	22
MOL-PCR_SNP	–	–	–			11	–	11[a]
Total	2	17	12	2	5	11	3	50 + 2

AFLP amplified fragment length polymorphism, *MOL-PCR* multiplex oligonucleotide ligation-PCR, *PC* positive control, *SGI1 Salmonella* genomic island 1, *SNP* single nucleotide polymorphism

[a]MOL-PCR_SNP includes 11 SNP alleles and 11 corresponding wild-type alleles and as such, 22 different regions of MagPlex-TAG beads are included in MOL-PCR_SNP

2 Materials

2.1 DNA Isolation

1. Deionized water, sterilized by autoclaving.
2. Luria Bertani (LB) agar plates, sterile.
3. Centrifuge with rotor for 1.5 mL microcentrifuge tubes, capable of centrifugation at 11,000 × *g*.
4. Heating block for 1.5 mL microcentrifuge tubes, capable of operating at 100 °C.
5. Incubator, capable of operating at 37 °C.
6. Micropipette 200 μL or smaller volume suitable for transfer of 50 μL.
7. Vortex.
8. 1 μL inoculation loops, disposable and sterile.
9. 1.5 mL microcentrifuge tubes, sterilized by autoclaving.
10. 0.5 mL microcentrifuge tubes, DNase-free or autoclaved.
11. 200 μL pipette tips, sterile and DNase-free or autoclaved.

2.2 Oligonucleotide Master and Working Stocks

1. Primer T7: 5′ TAATACGACTCACTATAGGG 3′.
2. Biotinylated primer T3: 5′ biotin- ATTAACCCTCACTAAAG GGA 3′.
3. Probes for MOL-PCR_1 (*see* Tables 2, 3, 4, and 5 [4]).
4. Probes for MOL-PCR_2 (*see* Tables 6, 7, and 8 [4]).
5. Probes for MOL-PCR_SNP (*see* Tables 9, 10, 11, and 12 [4]).
6. DNase/RNase-free distilled water, stored at 2–8 °C as single-use 1.5 mL aliquots.
7. Frozen cooling block for 0.5 mL microcentrifuge tubes.
8. Ice or frozen cooling block for 1.5–2 mL microcentrifuge tubes.
9. Micropipettes.
10. Minicentrifuge with rotor for 0.5 mL and 1.5–2 mL microcentrifuge tubes.
11. Vortex.
12. 0.5 mL microcentrifuge tubes, DNase-free or autoclaved.
13. Pipette tips, DNase-free or autoclaved.

2.3 Multiplex Oligonucleotide Ligation

1. Aliquot of DNA lysate of positive control *S.* Typhimurium (*see* **Note 1**), prepared as described in Subheading 3.1 DNA Isolation.
2. DNA lysate of samples, i.e., *S.* Typhimurium isolates to be characterized, prepared as described in Subheading 3.1 DNA Isolation.

Table 2
Probes included in probe premix 1.1 for MOL-PCR_1

Target	Targeted sequence	Probe	Sequence (5′ → 3′)	Bead region
All *Salmonella* species	*invA*	invA-U	*TAATACGACTCACTATAGGG*gataagaaagtgaaatgtaaattgATAAACTTCATCGCACCGTCA	51
		invA-D	P-AAGGAACCGTAAAGCTGGCTT*TCCCTTTAGTGAGGGTTAAT*	
Gifsy-1	*gipA*	SAL-29-U	*TAATACGACTCACTATAGGG*tttaagtgagttatagaagtagtaGGCAAGCTGTACATGGCAAAG	29
		SAL-29-D	P-AACAAAATCCCCTTAGACG*TCCCTTTAGTGAGGGTTAAT*	
Salmonella genomic island 1 (SGI1)	Left junction	SAL-50-U	*TAATACGACTCACTATAGGG*gtttgtgtttgtataagttgttaaTCTGCTTGTGTCTTTGGGT	53
		SAL-50-D	P-TCTCGTAGAGATAGAGTTCTAAAGG*TCCCTTTAGTGAGGGTTAAT*	
Salmonella genomic island 1 (SGI1)	Right junction	SAL-51-U	*TAATACGACTCACTATAGGG*tagagaaagagagaattgtattaaTGACGAGCTGAAGCGAATTG	54
		SAL-51-D	P-CAGGGCTGAACAGCTCAA*TCCCTTTAGTGAGGGTTAAT*	
MLVA locus STTR10	PSLT064	SAL-74-U	*TAATACGACTCACTATAGGG*tatgaatgttattgtgtgttgattCGGGCGCGGCTGGAGTATTTG	48
		SAL-74-D	P-CGCAACTCCCGGACAAGAAT*TCCCTTTAGTGAGGGTTAAT*	

Primer, anti-TAG, and target-specific sequences are indicated by, respectively, *italic*, *lower-case*, and *underlined* sequences. *MLVA* multiple-locus variable-number of tandem repeats analysis, *P* phosphate

Table 3
Probes included in probe premix 1.2 for MOL-PCR_1

Target	Targeted sequence	Probe	Sequence (5′ → 3′)	Bead region
S. Typhimurium	*rpoB*	rpoB-U	*TAATACGACTCACTATAGGG*gtaattgaattgaaagataagtgtTTTCTCAGCTGCACCGTAGC	18
		rpoB-D	P-CCTGGCGTCTTCTTTGACTCC*TCCCTTTAGTGAGGGTTAAT*	
ST64T/P22	g9–3’	SAL-11-U	*TAATACGACTCACTATAGGG*gtaagagtattgaaattagtaagaAACCCTTCTTCTGCCAAATTAGC	66
		SAL-11-D	P-AACATTAATTCTAGAGGGGTCTACC*TCCCTTTAGTGAGGGTTAAT*	
ST64T	*mnt*	SAL-16-U	*TAATACGACTCACTATAGGG*tgtatatgttaatgagatgttgtaTACAGGCGTAGGCTTTTCAAGTG	37
		SAL-16-D	P- AATCAGCCACTATCTGCACT*TCCCTTTAGTGAGGGTTAAT*	
ST64T/P22	g8	SAL-23-U	*TAATACGACTCACTATAGGG*agtagaaagttgaaattgattatgTGCCAACGACAATGCAGGTC	12
		SAL-23-D	P-AGGAAGAGGGCTTTGAGATTG*TCCCTTTAGTGAGGGTTAAT*	
AFLP fragments	GA-1	SAL-45-U	*TAATACGACTCACTATAGGG*gttagttatgatgaatattgtgtaGCGTGGAATATCGTTGATGG	45
		SAL-45-D	P-GTTTGAAGCGGTGTCGAA*TCCCTTTAGTGAGGGTTAAT*	
Allantoinase	*allB*	SAL-49-U	*TAATACGACTCACTATAGGG*agtaagtgttagatagtattgaatTTTCGCGACGTTAATGACT	38
		SAL-49-D	P-GGCAGTTTTACAAAGGCGCGC*TCCCTTTAGTGAGGGTTAAT*	
Gifsy-1	*artA*	SAL-53-U	*TAATACGACTCACTATAGGG*aaataagaatagagagagaaagttTCTGGTTATGCAAGTGCTGT	43
		SAL-53-D	P-TGATTTTGTATATCGTGTTGACTCGAG*TCCCTTTAGTGAGGGTTAAT*	
ST104B	*hldD* homolog	SAL-55-U	*TAATACGACTCACTATAGGG*aatgtaaagtaaagaaagtgatgaCGCAGTAGAGACATGGATGTA	44
		SAL-55-D	P-CCAAGTTGCACAGCGAAC*TCCCTTTAGTGAGGGTTAAT*	

Primer, anti-TAG, target-specific sequences and SNP positions are indicated by, respectively, *italic*, *lower-case*, *underlined*, and *boldface* sequences. *AFLP* amplified fragment length polymorphism, *P* phosphate, *SNP* single nucleotide polymorphism

Table 4
Probes included in probe premix 1.3 for MOL-PCR_1

Target	Targeted sequence	Probe	Sequence (5′ → 3′)	Bead region
Chloramphenicol/ florfenicol resistance	*floR*	SAL-67-U	*TAATACGACTCACTATAGGG*ttgtgatagtagttagatatttgtGCGGAATATTCCGA GATCGGA TTCAGC	39
		SAL-67-D	P-TTTGCCTTCGCCACTGTCGCGC*TCCCTTTAGTGAGGGTTAAT*	
Ampicillin resistance	*bla*$_{PSE\text{-}1}$	SAL-69-U	*TAATACGACTCACTATAGGG*gtgattgaatagtagattgtttaaCCCAATAGTAC AGTCGAGATT	46
		SAL-69-D	P-AAGAAAGCAGATCTTGTGACC*TCCCTTTAGTGAGGGTTAAT*	
Sulfonamide resistance	*sul1*	SAL-70-U	*TAATACGACTCACTATAGGG*gttgtaaattgtagtaaagaagtaCCCCAACGCCG ACTTCAGCTTT	15
		SAL-70-D	P-TGAAGGTTCGACAGCACGTGC*TCCCTTTAGTGAGGGTTAAT*	

Primer, anti-TAG, and target-specific sequences are indicated by, respectively, *italic*, *lower-case*, and *underlined* sequences. *P* phosphate

Table 5
Additional probes included in MOL-PCR_1

Target	Targeted sequence	Probe	Sequence (5′ → 3′)	Bead region
ST64T/P22	g9–5’	SAL-10-U	*TAATACGACTCACTATAGGG*gtgtgttatttgtttgtaaagtatGGCTGAGCAATCTGGACGTG	19
		SAL-10-D	P-AGAGCCATCCTCATTTTCAA*TCCCTTTAGTGAGGGTTAAT*	
ST104	g9	SAL-25-U	*TAATACGACTCACTATAGGG*gaagatattgaaagaatttgatgtCGAAGTTGGTCTGCACAATG	55
		SAL-25-D	P-CAGCAGGGCGTCCGTCGCTTC*TCCCTTTAGTGAGGGTTAAT*	
AFLP fragments	CA-1	SAL-33-U	*TAATACGACTCACTATAGGG*gttgagaattagaatttgataaagTTAACGATCAACGAAATTCAATCC	73
		SAL-33-D	P-TAACCACAGCCGCTGAAAAGGTT*TCCCTTTAGTGAGGGTAAT*	
Tetracycline resistance	*tet*(G)	SAL-68-U	*TAATACGACTCACTATAGGG*gattgatatttgaatgtttgtttgGAACGGTTGGGTTTGGATTGTCG	22
		SAL-68-D	P-GCGCGATCCTCTATTTAATATGTCTGCC*TCCCTTTAGTGAGGGTTAAT*	

Primer, anti-TAG, and target-specific sequences are indicated by, respectively, *italic*, *lower-case*, and ***underlined*** sequences. *AFLP* amplified fragment length polymorphism; *P* phosphate

Table 6
Probes included in probe premix 2.1 for MOL-PCR_2

Target	Targeted sequence	Probe	Sequence (5′ → 3′)	Bead region
ST64T/P22	*gtrA/gtrB*	SAL-14-U	*TAATACGACTCACTATAGGG*atttgttatgataaatgtgtagtgCTTTCTCGGCAATTAGCCTGGTATG	42
		SAL-14-D	P-CGGCTTTATCTATTCCAGAT*TCCCTTTAGTGAGGGTTAAT*	
ST64T/P22	*gtrC*	SAL-15-U	*TAATACGACTCACTATAGGG*agagtattagtagttattgtaagtGCATTAACACCTCTGACCACATC	57
		SAL-15-D	P-CAATTATTGTTAATAATGCGTGG*TCCCTTTAGTGAGGGTTAAT*	
P22	*sieB*	SAL-20-U	*TAATACGACTCACTATAGGG*tttgttagaatgagaagatttatgACAACTCATGGTGGCAGGAG	75
		SAL-20-D	P-CTAATGCGTTTTTTCCTGCA*TCCCTTTAGTGAGGGTTAAT*	
ST64T/P22	*eac*	SAL-21-U	*TAATACGACTCACTATAGGG*aaagaattagtatgatagatgagaCAGCTCTTTGTTGTATGCGC	76
		SAL-21-D	P-GGCCTTCCTTTGTGTTTCCC*TCCCTTTAGTGAGGGTTAAT*	
P22	g13	SAL-24-U	*TAATACGACTCACTATAGGG*tattagagtttgagaataagtagtCAGCGAAGCGTTTGATTAG	33
		SAL-24-D	P-CGAACCAATCGAGTCTGTG*TCCCTTTAGTGAGGGTTAAT*	
ST104	g44	SAL-26-U	*TAATACGACTCACTATAGGG*tttgatttaagagtgttgaatgtaCAACGCCCCACACACCA	26
		SAL-26-D	P-GGTTCGGTACCACCTTTAATG*TCCCTTTAGTGAGGGTTAAT*	
AFLP fragments	CA-3	SAL-35-U	*TAATACGACTCACTATAGGG*aataagagaattgatatgaagatgAATGGCTGGCAGGGTCTGTTC	35
		SAL-35-D	P-GAACCTGACGGACAGGCG*TCCCTTTAGTGAGGGTTAAT*	

Primer, anti-TAG, and target-specific sequences are indicated by, respectively, *italic*, *lower-case*, and ***underlined*** sequences. *AFLP* amplified fragment length polymorphism, *P* phosphate

Table 7
Probes included in probe premix 2.2 for MOL-PCR_2

Target	Targeted sequence	Probe	Sequence (5′ → 3′)	Bead region
S. Typhimurium	*rpoB*	rpoB-U	*TAATACGACTCACTATAGGG*gtaattgaattgaaagataagtgtTTTCTCAGCTGCACCGTAGC	18
		rpoB-D	P-CCTGGCGTCTTCTTTGACTCC*TCCCTTTAGTGAGGGTTAAT*	
ST104	g61	SAL-27-U	*TAATACGACTCACTATAGGG*aattgagaaagagataaatgatagCGCTACAGCAACAAAAACG	72
		SAL-27-D	P-TATGCTCCAGATGGAAGAGAGG*TCCCTTTAGTGAGGGTTAAT*	
AFLP fragments	CA-7	SAL-36-U	*TAATACGACTCACTATAGGG*attgtgaaagaaagagaagaaattCATCGGGGTGATCAGCA	14
		SAL-36-D	P-TGATTACCCTGTTTGCCCG*TCCCTTTAGTGAGGGTTAAT*	
AFLP fragments	CA26.1	SAL-37-U	*TAATACGACTCACTATAGG*gtatagtgtgattagatttgtaaaGCGGCGCGTATTTCGTGC	78
		SAL-37-D	P-ATCGCATTGTCATTACCAGCA*TCCCTTTAGTGAGGGTTAAT*	
AFLP fragments	CA28.4	SAL-38-U	*TAATACGACTCACTATAGGG*attaagtaagaattgagagtttgaTGATGCACTAATTCGTCG	21
		SAL-38-D	P-CGAGGCTGCTGGATATATTC*TCCCTTTAGTGAGGGTTAAT*	
AFLP fragments	CG-1	SAL-39-U	*TAATACGACTCACTATAGGG*aaattagttgaaagtatgagaaagCAGGGAACCGTCTTGAG	20
		SAL-39-D	P-CAAGTTCAGAGCGCAATGAC*TCCCTTTAGTGAGGGTTAAT*	
AFLP fragments	CT-2	SAL-43-U	*TAATACGACTCACTATAGGG*gtatgttgtaatgtattaagaaagCCCAGGTAAACAGGAAATCCA	25
		SAL-43-D	P-TTCCGGCACTGAAAATACTGC*TCCCTTTAGTGAGGGTTAAT*	
AFLP fragments	GA27.1	SAL-47-U	TAATACGACTCACTATAGGGaagatgatagttaagtgtaagttaGGTTTGTCCTACGACCCC	27
		SAL-47-D	P-GGAATCCTTCCATCGGAAATGA*TCCCTTTAGTGAGGGTTAAT*	
AFLP fragments	GC-1	SAL-48-U	*TAATACGACTCACTATAGGG*tgagtaagtttgtatgtttaagtaTGGAAGAACAAGCAAACAAGATTC	65
		SAL-48-D	P-TCGTAGAACTACTGCAAAAAGCCAG*TCCCTTTAGTGAGGGTTAAT*	

Primer, anti-TAG, target-specific sequences and SNP positions are indicated by, respectively, *italic*, *lower-case*, *underlined*, and *boldface* sequences. *AFLP* amplified fragment length polymorphism, *P* phosphate, *SNP* single nucleotide polymorphism

Table 8
Additional probes included in MOL-PCR_2

Target	Targeted sequence	Probe	Sequence (5′ → 3′)	Bead region
ST64T/P22	g17	SAL-12-U	*TAATACGACTCACTATAGGG*gtaagattagaagttaatgaagaaGCGTTGCAGTAATCAGGTTTG	52
		SAL-12-D	P-ATGTTGTTGTCATACTGAAATGC*TCCCTTTAGTGAGGGTTAAT*	
ST64T/P22	*int*	SAL-18-U	*TAATACGACTCACTATAGGG*aatgaaatagtgttaaatgagtgtCGGCAATCATCACAAATGG	74
		SAL-18-D	P-GTGTTCGTCTACAAGGAAAGC*TCCCTTTAGTGAGGGTTAAT*	
AFLP fragments	CG-2	SAL-40-U	*TAATACGACTCACTATAGGG*tattagagagaaattgtagagattCGCTATATCACGCTGAACAGAC	61
		SAL-40-D	P-CAGAACTAGGTGAAACAAAAGAGG*TCCCTTTAGTGAGGGTTAAT*	
AFLP fragments	CT-1	SAL-42-U	*TAATACGACTCACTATAGGG*tattgttgaatgtgtttaaagagaCATCTGCTGATAGCTTAGTTGTC	47
		SAL-42-D	P-GATAATGCCAACGACAATGCAGG*TCCCTTTAGTGAGGGTTAAT*	
Streptomycin/ spectinomycin resistance	*aadA2*	SAL-66-U	*TAATACGACTCACTATAGGG*ttgtgtagttaagagttgtttaatGGTATCTTCGAGCCAGCCAT	36
		SAL-66-D	P-GATCGACATTGATCTAGCTA*TCCCTTTAGTGAGGGTTAAT*	
Phase 2 flagellar gene	*fljB*	SAL-73-U	*TAATACGACTCACTATAGGG*tgaaatgtgtatttgtatgtttagCCAGCCGCAAGGGTTACTGTAC	62
		SAL-73-D	P-CGTCAGTAGCAACGTTAACTTCATAA*TCCCTTTAGTGAGGGTTAAT*	

Primer, anti-TAG, and target-specific sequences are indicated by, respectively, *italic*, *lower-case*, and *underlined* sequences. *AFLP* amplified fragment length polymorphism, *P* phosphate

Table 9
Probes included in probe premix S.1 for MOL-PCR_SNP

SNP	Targeted sequence	Probe	Sequence (5′ → 3′)	Bead region
TM81	STM0080	SAL-56-U_SNP	*TAATACGACTCACTATAGG*aattagaagtaagtagagtttaagGGTCGATAACGATACCGGTATGA	56
		SAL-56-D	P-TGTCTTACAAAAACGCCGCC*TCCCTTTAGTGAGGGTTAAT*	
TM1231	STM1269	SAL-59-U_SNP	*TAATACGACTCACTATAGG*tttgttgttaagtatgtgatttagCCAGCGCTTGATGGTCGGCT	63
		SAL-59-D	P-CTTTCAGCCACGACGATATGGTG*TCCCTTTAGTGAGGGTTAAT*	
TM973	STM1002	SAL-60-U_SNP	*TAATACGACTCACTATAGG*atgatgtgtttgatttgaattgaaGCTCATGTTCTGCTTATTAGC ACCGAAGGT	64
		SAL-60-D	P-GATACGTCCCCAGATATTTATGAAGATATAG*TCCCTTTAGT GAGGGTTAAT*	
1880	*otsA*	SAL-61-U_SNP	*TAATACGACTCACTATAGG*tttgtgtgttattgtaattgagatCACGATTGAGCGCCGCA	67
		SAL-61-D	P-GCCACGTCATCCCGATCG*TCCCTTTAGTGAGGGTTAAT*	
2199	*napF*	SAL-62-U_SNP	*TAATACGACTCACTATAGG*agtgaatgtaagattatgtatttgCGGAAGAAAAAGCGATTCA	13
		SAL-62-D	P-GTACAGGCCTGCGCACAG*TCCCTTTAGTGAGGGTTAAT*	
101	*araD*	SAL-63-U_SNP	*TAATACGACTCACTATAGG*attaagtaagaattgagagtttgaGGCGCGAATGGGTGTGTACA	21
		SAL-63-D	P-ATGCCGCCGATAGTCGG*TCCCTTTAGTGAGGGTTAAT*	
TM3124_1	intergenic	SAL-64-U_SNP	*TAATACGACTCACTATAGG*gatagatttagaatgaattaagtgTCACAACTTCAAAATAAAACGTTA TAAATTAATAT	28
		SAL-64-D	P-ATTATATCAACAATCGCTTTTATCCTTGC*TCCCTTTAGTGAGG GTTAAT*	
TM3275_2	intergenic	SAL-65-U_SNP	*TAATACGACTCACTATAGG*agtagaaagttgaaattgattatgCCTGCAAATACTCGTACGGG TCGCGC	12
		SAL-65-D	P-TCTTTTACATCATTACGACGTCAAAC*TCCCTTTAGTGAGGGTTAAT*	
4213	*yjeF*	SAL-71-U_SNP	*TAATACGACTCACTATAGG*gtatgttgtaatgtattaagaaagTCTGGCGATTATTGGCGGT	25
		SAL-71-D	P-GACCAGGGAACAGCGGGCGC*TCCCTTTAGTGAGGGTTAAT*	

Primer, anti-TAG, target-specific sequences and SNP positions are indicated by, respectively, *italic*, lower-case, underlined, and **boldface** sequences. *P* phosphate, *SNP* single nucleotide polymorphism

Table 10
Probes included in probe premix S.2 for MOL-PCR_SNP

SNP	Targeted sequence	Probe	Sequence (5′ → 3′)	Bead region
TM81	STM0080	SAL-56-U_WT	*TAATACGACTCACTATAGGG*tttgatttaagagtgttgaatgtaGGTCGATAACGATACCGGTATGG	26
TM1231	STM1269	SAL-59-U_WT	*TAATACGACTCACTATAGGG*tattagagtttgagaataagtagtCCAGCGCTTGATGGTCGGCG	33
TM973	STM1002	SAL-60-U_WT	*TAATACGACTCACTATAGGG*attgtgaaagaaagagaagaaattCTGCTTATTAGCACCGAAGGC	14
1880	*otsA*	SAL-61-U_WT	*TAATACGACTCACTATAGGG*tatgaatgttattgtgtgttgattCACGATTGAGCGCCGCC	48
2199	*napF*	SAL-62-U_WT	*TAATACGACTCACTATAGGG*agagtattagtagttattgtaagtGAAGAAAAAGCGATTCG	57
101	*araD*	SAL-63-U_WT	*TAATACGACTCACTATAGGG*aattgagaaagagataaatgatagCGAATGGGTGTGTACG	72
TM3124_1	intergenic	SAL-64-U_WT	*TAATACGACTCACTATAGGG*aatgaaatagtgttaaatgagtgtTCACAACTTCAAAATAAAACG TTATAAATTAATAG	74
TM3275_2	intergenic	SAL-65-U_WT	*TAATACGACTCACTATAGGG*tttgttagaatgagaagatttatgCCTGCAAATACTCGTACGGGT CGCGG	75
4213	*yjeF*	SAL-71-U_WT	*TAATACGACTCACTATAGGg*tttgtgtttgtataagttgttaaTCTGGCGATTATTGGCGGC	53

Primer, anti-TAG, target-specific sequences and SNP positions are indicated by, respectively, *italic*, *lower-case*, ***underlined***, and ***boldface*** sequences. *P* phosphate, *SNP* single nucleotide polymorphism

Table 11
Probes included in probe premix S.3 for MOL-PCR_SNP

SNP	Targeted sequence	Probe	Sequence (5′ → 3′)	Bead region
TM3230_1	intergenic	SAL-58-U_WT	*TAATACGACTCACTATAGG*gtgtgttatttgtttgtaaagtatGGCTTTGACGAAGACTAAACCT	19
		SAL-58-D	P-CCCCCTCTATACCCTATTTC*TCCCTTTAGTGAGGGTTAAT*	
TM2079	intergenic	SAL-72-U_SNP	*TAATACGACTCACTATAGGG*aatgtaaagtaaagaaagtgatgaTACATCAGGCAACGGTACGT	44
		SAL-72-U_WT	*TAATACGACTCACTATAGGG*atttgttatgataaatgtgtagtgATACATCAGGCAACGGTACGA	42
		SAL-72-D	P-CTATAGGACACCGCGCTAAG*TCCCTTTAGTGAGGGTTAAT*	

Primer, anti-TAG, target-specific sequences and SNP positions are indicated by, respectively, *italic*, *lower-case*, *underlined*, and *boldface* sequences. *P* phosphate, *SNP* single nucleotide polymorphism

Table 12
Additional probes included in MOL-PCR_SNP

SNP	Targeted sequence	Probe	Sequence (5′ → 3′)	Bead region
TM3230_1	intergenic	SAL-58-U_SNP	*TAATACGACTCACTATAGGG*ttgtgatagtagttagatatttgtGGCTTTGACGAAGACTAAACCC	39

Primer, anti-TAG, target-specific sequences and SNP positions are indicated by, respectively, *italic*, *lower-case*, *underlined*, and ***boldface*** sequences. *P* phosphate, *SNP* single nucleotide polymorphism

3. DNase/RNase-free distilled water, stored at 2–8 °C as single-use 1.5 mL aliquots.
4. Aliquot of working stocks at 5 μM of probe premixes 1.1, 1.2, 1.3, 2.1, 2.2, S.1, S.2, S.3 and of additional probes for MOL-PCR_1, MOL-PCR-2, and MOL-PCR_SNP, prepared as described in Subheading 3.2 Oligonucleotide Master and Working Stocks.
5. *Taq* DNA ligase enzyme 40 U/μL.
6. Aliquot of *Taq* DNA ligase buffer 10×.
7. Ice.
8. Frozen cooling blocks for 0.5 mL microcentrifuge tubes.
9. Frozen cooling blocks for 1.5–2 mL microcentrifuge tubes.
10. Frozen cooling blocks for PCR tubes, strips, or plate.
11. Micropipettes 2, 20, 200, 1000 μL.
12. Multichannel pipette 20 μL or smaller volume suitable for transfer of 3 μL.
13. Minicentrifuge with rotor for 0.5 mL and 1.5 mL microcentrifuge tubes.
14. Minicentrifuge for PCR plates or with rotor for PCR tubes.
15. Thermal cycler (*see* **Note 2**).
16. Vortex.
17. 1.5 mL microcentrifuge tubes, DNase-free or autoclaved.
18. PCR tubes, strips, or plates with compatible sealing.
19. Pipette tips, DNase-free or autoclaved.

2.4 Singleplex PCR with Universal Primers

1. Ligation products, prepared as described in Subheading 3.3 Multiplex Oligonucleotide Ligation.
2. Aliquot of working stock at 20 μM of primers T7 and biotinylated T3, prepared as described in Subheading 3.2 Oligonucleotide Master and Working Stocks.
3. DNase/RNase-free distilled water, stored at 2–8 °C as single-use 1.5 mL aliquots.
4. Hot start *Taq* DNA polymerase.
5. Hot start *Taq* DNA polymerase buffer 10×.
6. Aliquot of dNTP mix 10 mM each.
7. Ice.
8. Frozen cooling blocks for 0.5 mL microcentrifuge tubes.
9. Frozen cooling blocks for 1.5–2 mL microcentrifuge tubes.
10. Frozen cooling blocks for PCR tubes, strips, or plate.
11. Micropipettes 2, 20, 200, 1000 μL.

12. Multichannel pipette 20 μL or smaller volume suitable for transfer of 3 μL.
13. Minicentrifuge with rotor for 0.5 mL and 1.5 mL microcentrifuge tubes.
14. Minicentrifuge for PCR plates or with rotor for PCR tubes.
15. Thermal cycler (*see* **Note 2**).
16. Vortex.
17. 1.5 mL microcentrifuge tubes, DNase-free or autoclaved.
18. PCR tubes, strips, or plates with compatible sealing.
19. Pipette tips, DNase-free or autoclaved.

2.5 Hybridization to MagPlex-TAG Beads and Incubation with SAPE

1. Aliquot of 1× Tm hybridization buffer: 0.1 M Tris–HCl pH 8.0, 0.2 M NaCl, 0.08% Triton X-100, DNase/RNase-free distilled water, sterilized by filtration (0.2 μM) and stored at 2–8 °C in 50 mL aliquots (*see* **Note 3**).
2. Aliquot of 1.25× Tm hybridization buffer: 0.125 M Tris–HCl pH 8.0, 0.25 M NaCl, 0.1% Triton X-100, DNase/RNase-free distilled water, sterilized by filtration (0.2 μM) and stored at 2–8 °C in 15 mL aliquots (*see* **Note 3**).
3. MOL-PCR products, prepared as described in Subheading 3.4 Singleplex PCR with Universal Primers.
4. MagPlex-TAG beads, concentration 2.5 × 10^6 beads/mL, *see* Table 2 up to Table 12 for the appropriate bead regions (*see* **Note 4**).
5. Aliquot of SAPE 1 mg/ml (*see* **Note 4**).
6. Bath sonicator.
7. MAGPIX instrument (*see* **Note 5**).
8. Magnetic particle concentrator for PCR plates (*see* **Note 6**).
9. Minicentrifuge for PCR plates or with rotor for PCR tubes.
10. Micropipettes 10, 20, 200, 1000, 5000 μL.
11. Multichannel micropipette 20, 200 μL.
12. Thermal cycler.
13. Vortex.
14. Ice.
15. Aluminum foil.
16. 15-mL centrifuge tubes, DNase-free.
17. Dark holder for PCR plates.
18. 1.5 mL or 2 mL amber microcentrifuge tubes, DNase-free or autoclaved.
19. Reagent reservoir for multichannel pipette, DNase-free.

20. PCR plate compatible with the MAGPIX instrument with cover (*see* **Note 7**).
21. Pipette tips, DNase-free or autoclaved.
22. Filter tips 200 μL, DNase-free.

3 Methods

Wear clean powder-free gloves during the molecular laboratory manipulations to protect the DNA lysate, reagents and MOL-PCR products from DNase degradation and contaminating DNA originating from other sources. The work bench and the pipettes should be cleaned using appropriate solutions (e.g., alcohol, Umonium, 5% Dettol) before and after use. Do not vortex concentrated enzyme solutions.

3.1 DNA Isolation

1. Prepare a fresh culture of the *S.* Typhimurium isolate by streaking a small amount of bacterial cells on a sterile Luria–Bertani agar plate (*see* **Note 8**).
2. Incubate the inoculated Luria–Bertani agar plate overnight (14–20 h) at 37 °C.
3. Prepare one 50 μL aliquot of sterile deionized water in a sterile 1.5 mL microcentrifuge tube for each *S.* Typhimurium isolate.
4. Pick 1 colony of the overnight culture with a sterile, disposable 1 μL inoculation loop. Release the bacterial cells into the deionized water by rotating the inoculation loop in the 1.5 mL microcentrifuge tube.
5. Close the 1.5 mL microcentrifuge tube and disperse any clumps by vortexing for 5–10 s at maximum speed.
6. Perform a heat lysis of the bacterial cells by incubating the 1.5 mL microcentrifuge tube for 10 min at 100 °C in a heating block.
7. Cool the 1.5 mL microcentrifuge tube for 5–15 min in a refrigerator at 2–8 °C.
8. Centrifuge 10 min at 11,000 × *g*, room temperature. The debris of the lysed bacterial cells will precipitate, while the DNA remains in solution in the supernatant.
9. Transfer the supernatant to a DNase-free 0.5 mL microcentrifuge tube without disturbing the pellet (*see* **Note 9**). Discard the 1.5 mL microcentrifuge tube containing the pellet.
10. Store the DNA lysate at −20 °C for maximum 6 months (*see* **Note 10**).

3.2 Oligonucleotide Master and Working Stocks

1. Prepare 200 μM master stock solutions of primers T7 and biotinylated T3 by adding the necessary volume of DNase/RNase-free distilled water. Vortex at medium speed and spin down in a minicentrifuge. Prepare 10 μL aliquots of the T7 master stock solution and 30 μL aliquots of the biotinylated T3 master stock solution in 0.5 mL microcentrifuge tubes. Store the aliquots for maximum 2 years at −20 °C.
2. Prepare 20 μM working stock solutions of primers T7 and biotinylated T3. Hereto, add, respectively, 90 μL and 270 μL of DNase/RNase-free distilled water to aliquots of 200 μM master stock solutions of primers T7 and biotinylated T3 which were thawed on ice. Vortex at medium speed and spin down in a minicentrifuge. Divide these 20 μM working stock solutions in 10 μL aliquots in 0.5 mL microcentrifuge tubes. Store the aliquots for maximum 2 years at −20 °C.
3. Prepare 100 μM master stock solutions of each upstream and downstream probe (*see* Tables 2, 3, 4, 5, 6, 7, 8, 9, 10, 11, and 12) by adding the necessary volume of DNase/RNase-free distilled water. Vortex at medium speed and spin down in a minicentrifuge. Prepare 10 μL aliquots of the 100 μM master stock solutions in 0.5 mL microcentrifuge tubes. Store the aliquots for maximum 2 years at −20 °C.
4. Prepare 5 μM working stock solutions for each probe premix (*see* Tables 2, 3, 4, 6, 7, 9, 10, and 11) (*see* **Note 11**). Thaw an aliquot of the 100 μM master stock solution of the necessary probes on ice and add 2 μL of each master stock solution to the required volume of cold DNase/RNase-free distilled water to a total volume of 40 μL in a 0.5 mL microcentrifuge tube. Vortex at medium speed and spin down in a minicentrifuge. Immediately divide the 5 μM working stock into 2 μL aliquots. Spin down the aliquots and store at −20 °C for maximum 2 months (*see* **Note 12**).
5. Prepare 5 μM working stock solutions for each additional probe (*see* Tables 5, 8, and 12) (*see* **Note 11**). Thaw an aliquot of the 100 μM master stock solution of the probe on ice and add 2 μL of the master stock solution to 38 μL of cold DNase/RNase-free distilled water in a 0.5 mL microcentrifuge tube. Vortex at medium speed and spin down in a minicentrifuge. Immediately divide the 5 μM working stock into 2 μL aliquots. Spin down the aliquots and store at −20 °C for maximum 2 months (*see* **Note 12**).

3.3 Multiplex Oligonucleotide Ligation

1. Let the DNA lysate of positive controls and samples thaw on ice. Vortex at medium speed and spin down in a minicentrifuge. Keep the DNA lysate on ice or in frozen cooling blocks.
2. Make the calculation for the required ligation mix according to Tables 13, 14, and 15. Take all samples, the positive controls, one negative control, and one reaction as overage into account.

Table 13
Ligation mix for MOL-PCR_1

Reagent	Working stock	Final concentration	Volume per reaction (μL)
DNase/RNase-free distilled water	–	–	2.55
Taq DNA ligase buffer	10×	1×	1
Probe mix 1.2	50 nM	2 nM	0.4
Probe mix 1.3	50 nM	2 nM	0.4
SAL-10-U	50 nM	2 nM	0.4
SAL-10-D	50 nM	2 nM	0.4
SAL-25-U	50 nM	2 nM	0.4
SAL-25-D	50 nM	2 nM	0.4
SAL-33-U	50 nM	2 nM	0.4
SAL-33-D	50 nM	2 nM	0.4
Probe mix 1.1	50 nM	2 nM	0.4
SAL-68-U	50 nM	2 nM	0.4
SAL-68-D	50 nM	2 nM	0.4
Taq DNA ligase	40 U/μL	0.2 U/μL	0.05
		Total volume	8

3. Preheat the thermal cycler to 95 °C.
4. Let an aliquot of *Taq* DNA ligase buffer thaw on ice. Keep aliquots of DNase/RNase free distilled water cold on ice.
5. Take the required 2 μL aliquots of probe premixes and additional probes at a concentration of 5 μM from the freezer and keep them in a frozen cooling block, also during transport. Dilute the probe premixes and additional probes to a working stock at a concentration of 50 nM by adding 198 μL of cold DNase/RNase-free distilled water to the 2 μL aliquot while keeping the microcentrifuge tube with the aliquot in the frozen cooling block. Mix the diluted probe mixes and additional probes by vortexing for 10 s at medium speed, spin down, vortex again for 10 s at medium speed and spin down. Place the microcentrifige tubes immediately back in the cooling block.
6. Prepare the required ligation mix as specified in Tables 13, 14, and 15 in a 1.5 mL microcentrifuge tube which is placed in a frozen cooling block. Pipet first the DNase/RNase-free distilled water, then the *Taq* DNA ligase buffer which is first mixed by 10 s vortexing at medium speed and spun down,

Table 14
Ligation mix for MOL-PCR_2

Reagent	Working stock	Final concentration	Volume per reaction (μL)
DNase/RNase-free distilled water	–	–	1.35
Taq DNA ligase buffer	10×	1×	1
Probe mix 2.1	50 nM	2 nM	0.4
Probe mix 2.2	50 nM	2 nM	0.4
SAL-12-U	50 nM	2 nM	0.4
SAL-12-D	50 nM	2 nM	0.4
SAL-18-U	50 nM	2 nM	0.4
SAL-18-D	50 nM	2 nM	0.4
SAL-40-U	50 nM	2 nM	0.4
SAL-40-D	50 nM	2 nM	0.4
SAL-66-U	50 nM	2 nM	0.4
SAL-66-D	50 nM	2 nM	0.4
SAL-73-U	50 nM	2 nM	0.4
SAL-73-D	50 nM	2 nM	0.4
SAL-42-U	50 nM	2 nM	0.4
SAL-42-D	50 nM	2 nM	0.4
Taq DNA ligase	40 U/μL	0.2 U/μL	0.05
		Total volume	8

Table 15
Ligation mix for MOL-PCR_SNP

Reagent	Working stock	Final concentration	Volume per reaction (μL)
DNase/RNase-free distilled water	–	–	5.35
Taq DNA ligase buffer	10×	1×	1
Probe mix S.1	50 nM	2 nM	0.4
Probe mix S.2	50 nM	2 nM	0.4
Probe mix S.3	50 nM	2 nM	0.4
SAL-58-U_SNP	50 nM	2 nM	0.4
Taq DNA ligase	40 U/μL	0.2 U/μL	0.05
		Total volume	8

followed by the working stock of probe mixes and additional probes. Just before pipetting the *Taq* DNA ligase enzyme, take the aliquot from the freezer and transport it on ice or in a frozen cooling block. Mix by tapping the tube and spin down, do not vortex. Add the *Taq* DNA ligase enzyme to the ligation mix and place the aliquot of *Taq* DNA ligase enzyme immediately back at −20 °C.

7. Take a new, frozen cooling block from the freezer and place herein PCR tubes or strips or a PCR plate.
8. Spin down the ligation mix in the 1.5 mL microcentrifuge tube and vortex for 10 s at medium speed. Repeat this step.
9. Spin down the ligation mix and add 8 μL of the ligation mix to each well for the negative control, the positive controls and the samples. Keep the PCR tubes, strips, or plate in the cooling block.
10. While keeping the PCR tubes, strips, or plate in the cooling block, add 2 μL of DNase/RNase-free distilled water to the well for the negative control, which is a no-template-control (NTC) (*see* **Note 13**). Add 2 μL of DNA lysate of each positive control and of each sample to the corresponding well (*see* **Note 9**). Pipet the DNase/RNase-free distilled water or DNA lysate into the ligation mix.
11. Cover and spin down the PCR tubes, strips, or plate and transport to the thermal cycler in a cooling block.
12. Perform the multiplex oligonucleotide ligation reaction according to following parameters: initial denaturation at 95 °C for 10 min, 30 cycles of 94 °C for 25 s and 58 °C for 30 s, final hold at 16 °C.
13. The ligation products can be stored up to 1 day at 2–8 °C or at −20 °C for longer storage.

3.4 Singleplex PCR with Universal Primers

1. If the singleplex PCR is done on frozen ligation products, let the ligation products thaw on ice. Keep on ice or in frozen cooling blocks.
2. Make the calculation for the required PCR mix according to Table 16. Take all samples, the positive controls, the negative control, and one reaction as overage into account.
3. Preheat the thermal cycler to 95 °C.
4. Let an aliquot of dNTP mix, of biotinylated T3 primer and of T7 primer thaw on ice. Thaw Hot start *Taq* DNA polymerase buffer.
5. Prepare the required PCR mix as specified in Table 16 in a 1.5 mL microcentrifuge tube which is placed on ice or in a frozen cooling block (*see* **Note 14**). Pipet first the DNase/RNase-free distilled water, then the Hot start *Taq* DNA polymerase

Table 16
Singleplex PCR mix

Reagent	Working stock	Final concentration	Volume per reaction (μL)
DNase/RNase-free distilled water	–	–	5.4375
Hot start *Taq* DNA polymerase buffer	10×	1×	1
T7 primer	20 μM	125 nM	0.0625
Biotinylated T3 primer	20 μM	500 nM	0.25
dNTP mix	10 mM	200 μM	0.2
Hot start *Taq* DNA polymerase	5 U/μL	0.025 U/μL	0.05
		Total volume	7

buffer which is mixed by vortexing 10 s at full speed and spun down, followed by the working stock of T7 and biotinylated T3 primer and the dNTP mix. Just before pipetting the Hot start *Taq* DNA polymerase enzyme, take the tube from the freezer and transport it on ice or in a frozen cooling block. Mix by tapping the tube and spin down, do not vortex. Add the Hot start *Taq* DNA polymerase enzyme to the PCR mix and place the tube with Hot start *Taq* DNA polymerase enzyme immediately back at −20 °C.

6. Take a new, frozen cooling block from the freezer and place herein PCR tubes or strips or a PCR plate.
7. Spin down the PCR mix in the 1.5 mL microcentrifuge tube and vortex for 10 s at medium speed, repeat this step.
8. Spin down the PCR mix and add 7 μL of the PCR mix to each well for the negative controls, the positive controls and the samples. Keep the PCR tubes, strips, or plate in the cooling block.
9. Spin down the PCR tubes, strips, or plate with the ligation products of the finished multiplex oligonucleotide ligation reaction.
10. While keeping the PCR tubes, strips, or plate in the cooling block, add 3 μL of the ligation product to the corresponding well of the PCR tubes, strips, or plate. Pipet the ligation product into the PCR mix. This step can be done with a multichannel pipette.
11. Cover the PCR tubes, strips, or plate and spin down.

12. Perform the singleplex PCR according to following parameters: activation and initial denaturation at 95 °C for 15 min (*see* **Note 15**), 35 cycles of 94 °C for 30 s, 60 °C for 30 s, and 72 °C for 30 s, final extension at 72 °C for 5 min and final hold at 16 °C.
13. The MOL-PCR products can be stored up to 1 day at 2–8 °C or at −20 °C for longer storage.

3.5 Hybridization to MagPlex-TAG Beads and Incubation with SAPE

1. If the MOL-PCR products are frozen, let them thaw on ice and keep them on ice.
2. Let an aliquot of 1.25× Tm hybridization buffer equilibrate to room temperature.
3. Remove the appropriate MagPlex-TAG beads from the refrigerator and allow the bead suspension to equilibrate to room temperature for 5 min.
4. Prepare the required bead mix by diluting the MagPlex-TAG beads to a concentration of 37.5 beads/μL in 1.25× Tm hybridization buffer. Prepare 22 μL (20 μL plus 10% overage) for each positive control, negative control, and sample. Hereto, pipet first the required volume of 1.25× Tm hybridization buffer into an amber microcentrifuge tube or into a 15 mL centrifuge tube covered with aluminum foil. Next, place the bottles with MagPlex-TAG beads on a vortex at low speed and gently rotate for 2 min. Immediately before pipetting the required volume, mix the MagPlex-TAG beads in the bottle by gentle inversion for ten times and gently tap the bottle on the bench to minimize suspension retention in the cap.
5. Preheat the thermal cycler to 96 °C.
6. Spin down the PCR tubes, strips, or plate with the MOL-PCR products.
7. Vortex the bead mix for 10 s at medium and sonicate for 10 s. Do not centrifuge as this will cause precipitation of the beads.
8. Add immediately 20 μL of bead mix to the required wells in the PCR plate compatible with the MAGPIX instrument. Place the PCR plate in a dark holder to protect the beads from light.
9. Add 5 μL of MOL-PCR product to the corresponding wells with a multichannel pipette. Pipet the MOL-PCR product into the bead mix. Do not mix by pipetting up and down as this will cause foaming.
10. Cover the PCR plate and mix MOL-PCR product with the bead mix by shifting the plate gently ten times back and forth from the left-most to the right most position of the magnetic particle concentrator.
11. Perform the hybridization in the preheated thermal cycler with following parameters: 96 °C for 90 s and 37 °C for 30 min.

12. Carry out the necessary maintenance (e.g., calibration, verification, and fluidics) on the MAGPIX instrument during the hybridization reaction.
13. Just before the hybridization reaction ends, prepare a solution of SAPE 4 μg/mL in 1× Tm hybridization buffer in an amber microcentrifuge tube or in a 15 mL centrifuge tube covered with aluminum foil. Prepare 110 μL (100 μL + 10% overage) for each positive control, negative control and sample.
14. When the hybridization reaction is finished, remove the PCR plate from the thermal cycler and place the PCR plate in a dark holder. Preheat the thermal cycler to 37 °C.
15. Gently remove the cover from the PCR plate. Add 100 μL of the SAPE solution to each well. To save time, use a multichannel pipette in reverse mode to avoid foaming. Use filter tips and do not touch the wells to avoid contamination.
16. Cover the PCR plate and mix by shifting the plate gently ten times back and forth from the left-most to the right most position of the magnetic particle concentrator.
17. Perform the SAPE incubation in the preheated thermal cycler at 37 °C for 15 min.
18. Analyze immediately on a MAGPIX instrument.

3.6 Analysis on MAGPIX

1. Create the required protocol for the MAGPIX. The acquisition settings should be set to a volume of 100 μL, the XY-heater enabled at 37 °C and the sample wash enabled. "None" should be selected as analysis type and minimum bead count should be set to 50.
2. Create a batch from the appropriate protocol. Take sample type "Unknown" for positive controls and samples, and sample type "Background" for the negative control (NTC).
3. Preheat the heater block of the MAGPIX to 37 °C.
4. When the SAPE incubation finishes, transfer the PCR plate to the heating block of the MAGPIX instrument and carefully remove the cover.
5. Check visually that every well contains the same volume, which should be 125 μL. If a well contains less volume, remove this well from the created batch. Check also for the presence of bubbles and remove them with a clean pipette tip if needed.
6. Retract the plate into the MAGPIX and start the analysis of the batch.

3.7 Data Interpretation

1. Calculate signal-to-noise ratios (SN) for each sample and each marker according to Eq. 1 from the median fluorescence intensity (MFI) values in the comma-separated values (csv) file which is given as output by the MAGPIX instrument (*see* **Notes 16** and **17**).

$$SN_{\text{Sample } x_{\text{marker } a}} = \frac{MFI_{\text{Sample } x_{\text{marker } a}}}{MFI_{\text{NTC}_{\text{marker } a}}} \tag{1}$$

2. For the markers in MOL-PCR_SNP, calculate an SNP allele call and a wild-type (WT) allele call from the signal-to-noise ratios (*SN*) according to Eqs. 2 and 3.

$$\text{Allele call} - \text{SNP}_{\text{Sample } x \text{ SNP } a} = \frac{SN_{\text{Sample } x_{\text{SNP } a}}}{SN_{\text{Sample } x_{\text{SNP a}}} + SN_{\text{Sample } x \text{ WT } a}} \tag{2}$$

$$\text{Allele call} - \text{WT}_{\text{Sample } x \text{ WT } a} = \frac{SN_{\text{Sample } x_{\text{WT } a}}}{SN_{\text{Sample } x_{\text{SNP a}}} + SN_{\text{Sample } x \text{ WT } a}} \tag{3}$$

3. Compare the signal-to-noise ratios for the markers in MOL-PCR_1 and MOL-PCR_2 to the corresponding cutoff values (*see* Tables 17 and 18) to determine if the sample is positive or negative for that marker (*see* **Note 18**). If the sample is positive for a marker, i.e., the signal-to-noise ratio is higher than the cutoff value, assign the corresponding prime number for that marker (*see* Tables 17 and 18) to the sample. If the sample is negative for a marker, i.e., the signal-to-noise ratio is lower than or equal to the cutoff value, assign the number "1" for that marker to the sample.
4. Compare the MFI values for the markers in MOL-PCR_SNP to the corresponding cutoff values (*see* Table 19) to determine if the SNP locus is present in the sample (*see* **Note 18**). If an SNP locus is absent, i.e., the MFI value of the SNP allele and of the wild-type allele are both lower than the corresponding cutoff value, assign the corresponding prime number for absence of that SNP locus (*see* Table 19) to the sample and go to **step 6**. If an SNP locus is present, i.e., the MFI value of the SNP allele or of the wild-type allele is higher than the corresponding cutoff value, go to **step 5**.
5. Compare the allele call values for MOL-PCR_SNP to the cutoff value 0.6 to determine if the SNP or the WT allele is present in the sample. If the SNP allele is present, i.e., the SNP allele call is higher than 0.6, assign the corresponding prime number for the presence of the SNP (*see* Table 19) to the sample. Otherwise, if the wild-type allele is present, i.e., the wild-type allele call is higher than 0.6, assign the number "1" for the marker to the sample.
6. Calculate the Gödel Prime Product (GPP) [8–10] by multiplying the assigned prime numbers or "1" for all markers of MOL-PCR_1, MOL-PCR_2, and MOL-PCR_SNP separately. As such, the result of the multiplex assay is summarized in a MOL-PCR profile that consists of 3 GPPs, i.e., $\text{GPP}_{\text{MOL-PCR\_1}}$–$\text{GPP}_{\text{MOL-PCR\_2}}$–$\text{GPP}_{\text{MOL-PCR\_SNP}}$ (*see* **Note 19**).

Table 17
Cutoff values for signal-to-noise ratios and assigned prime numbers for markers in MOL-PCR_1

Marker	Cutoff value	Prime number
invA	8	2
rpoB	10	2
SAL-10	3.5	7
SAL-11	5	13
SAL-16	3.5	17
SAL-23	6	11
SAL-25	6	31
SAL-29	3.5	5
SAL-33	3.5	67
SAL-45	4	19
SAL-49	5	3
SAL-50	4	43
SAL-51	4	47
SAL-53	5	37
SAL-55	3.5	41
SAL-67	6	53
SAL-68	3.5	61
SAL-69	5	59
SAL-70	7	29
SAL-74	6	23

4 Notes

1. DNA lysate of *S.* Typhimurium isolates with a known MOL-PCR profile is used as positive control for the reaction. *S.* Typhimurium isolates for positive control should be chosen so that a positive result is expected for each of the markers in the MOL-PCR assay.
2. We have experienced that a thermal cycler with a maximum heating rate of 3.3 °C/s and cooling rate of 2.0 °C/s resulted in higher MFI values for the no-template control than a thermal cycler with a maximum heating and cooling rate of 4 °C/s [5].

Table 18
Cutoff values for signal-to-noise ratios and assigned prime numbers for markers in MOL-PCR_2

Marker	Cutoff value	Prime number
rpoB	4	2
SAL-12	4	41
SAL-14	3	23
SAL-15	3	7
SAL-18	3	11
SAL-20	3	17
SAL-21	3	53
SAL-24	3	59
SAL-26	3	29
SAL-27	3	13
SAL-35	3	19
SAL-36	8	71
SAL-37	4	61
SAL-38	3.5	67
SAL-39	5	37
SAL-40	3	73
SAL-42	3	5
SAL-43	5	79
SAL-47	3	47
SAL-48	3.5	31
SAL-66	5	43
SAL-73	6	3

3. All components of Tm hybridization buffer should be suitable for molecular biology.
4. MagPlex-TAG beads and streptavidin-R-phycoerythrin (SAPE) should be protected from light.
5. We have used a MAGPIX instrument, but other Luminex platforms may be used. However, details for creating a protocol and a batch depend on the Luminex platform.
6. A magnetic particle concentrator which pulls the beads to the side of each well and which has 13 columns is required to be able to mix the samples with the beads without causing foam due to the detergent in the hybridization buffer.

Table 19
Cutoff values for MFI value and assigned prime numbers for SNPs and presence of the SNP locus for markers in MOL-PCR_SNP

Marker	Cutoff value	Prime number presence of SNP	Prime number absence of locus
SAL-56_SNP	600	5	47
SAL-56_WT	600	1	
SAL-58_SNP	500	29	41
SAL-58_WT	700	1	
SAL-59_SNP	350	11	53
SAL-59_WT	350	1	
SAL-60_SNP	1000	3	59
SAL-60_WT	350	1	
SAL-61_SNP	250	13	61
SAL-61_WT	800	1	
SAL-62_SNP	350	19	67
SAL-62_WT	200	1	
SAL-63_SNP	700	17	71
SAL-63_WT	400	1	
SAL-64_SNP	700	31	73
SAL-64_WT	200	1	
SAL-65_SNP	800	7	79
SAL-65_WT	400	1	
SAL-71_SNP	1200	23	43
SAL-71_WT	350	1	
SAL-72_SNP	1200	37	83
SAL-72_WT	250	1	

7. The cover should be easily removed from the PCR plate.
8. *Salmonella* cultures should be handled in a laboratory at biosafety level 2. Work in a laminar flow hood or in the sterile area of a burning Bunsen burner while handling *Salmonella* cultures to prevent contamination of the cultures.
9. If the DNA isolation has to be performed on many *S.* Typhimurium isolates, it may be convenient to collect the DNA lysate in a DNase-free 96-well microtiter plate with adequate sealing for storage at −20 °C. As such, the DNA lysate can be transferred to the multiplex oligonucleotide reaction with a

multichannel pipette. Leave enough wells available in the first column for the positive and negative controls of the MOL-PCR assay.

10. As the isolated DNA is not purified, it should be kept at −20 °C or on ice to prevent DNase activity. The DNA lysate of positive control *S.* Typhimurium isolates should be stored in 10 μL aliquots, to avoid repetitive freeze–thaw cycles.
11. Probe working stock solutions should remain cold at all times. It is thus necessary to place all microcentrifuge tubes on a frozen cooling block and to replace regularly the cooling block with a new, frozen cooling block. Keep also the used DNase/RNase-free distilled water on ice or on a frozen cooling block.
12. These aliquots are single-use only to prevent repetitive freeze–thaw cycles and should be discarded immediately after use.
13. A no-template-control (NTC) is required for the MOL-PCR assay to check for contaminations and to measure background fluorescence intensity.
14. The PCR mix can be prepared before the multiplex oligonucleotide ligation reaction is finished to save time.
15. Depending on the type of Hot start *Taq* DNA polymerase used, the activation step can be different. The initial denaturation should at least be performed at 95 °C for 10 min.
16. R software [11] is recommended for these calculations.
17. A R Shiny application [12] is available for performing all calculations from the csv output files of the MAGPIX instrument, and which returns the GPPs and a graphical representation of all MOL-PCR profiles present in the analyzed *S.* Typhimurium isolates [4].
18. The cutoff values in Tables 17, 18, and 19 were calculated after applying the MOL-PCR assay on 519 *S.* Typhimurium and *S.* 1,4,[5],12:i:- isolates [4]. They may require adjustment according to the Luminex instrument used for performing the analysis.
19. $GPP_{MOL\text{-}PCR\_SNP}$ is calculated with the prime numbers or "1" for the absence or presence of the SNP locus (based on MFI values) and with the prime numbers or "1" for presence or absence of the SNP (based on allele call).

Acknowledgment

This work was supported by grant P4044.0103 (SalMolType) from the Scientific Institute of Public Health (WIV-ISP – RP/PJ). The National Reference Centre for *Salmonella* and *Shigella* is partially supported by the Belgian Ministry of Social Affairs through a fund within the Health Insurance System.

References

1. Deshpande A, Gans J, Graves SW, Green L, Taylor L, Kim HB, Kunde YA, Leonard PM, Li P-E, Mark J, Song J, Vuyisich M, White PS (2010) A rapid multiplex assay for nucleic acid-based diagnostics. J Microbiol Methods 80(2):155–163. doi:10.1016/j.mimet.2009.12.001
2. Stucki D, Malla B, Hostettler S, Huna T, Feldmann J, Yeboah-Manu D, Borrell S, Fenner L, Comas I, Coscollà M, Gagneux S (2012) Two new rapid SNP-typing methods for classifying *Mycobacterium tuberculosis* complex into the main phylogenetic lineages. PLoS One 7(7):e41253. doi:10.1371/journal.pone.0041253
3. Thierry S, Hamidjaja RA, Girault G, Löfström C, Ruuls R, Sylviane D (2013) A multiplex bead-based suspension array assay for interrogation of phylogenetically informative single nucleotide polymorphisms for *Bacillus anthracis*. J Microbiol Methods 95(3):357–365. doi:10.1016/j.mimet.2013.10.004
4. Wuyts V, Mattheus W, Roosens NHC, Marchal K, Bertrand S, De Keersmaecker SCJ (2015) A multiplex oligonucleotide ligation-PCR as a complementary tool for subtyping of *Salmonella* Typhimurium. Appl Microbiol Biotechnol 99(19):8137–8149. doi:10.1007/s00253-015-6831-7
5. Wuyts V, Roosens NHC, Bertrand S, Marchal K, De Keersmaecker SCJ (2015) Guidelines for optimisation of a multiplex oligonucleotide ligation-PCR for characterisation of microbial pathogens in a microsphere suspension array. BioMed Res Int 2015:790170. doi:10.1155/2015/790170
6. European Food Safety Authority (EFSA), European Centre for Disease Prevention and Control (ECDC) (2015) The European Union summary report on trends and sources of zoonoses, zoonotic agents and food-borne outbreaks in 2013. EFSA J 13(1):3991. doi:10.2903/j.efsa.2015.3991
7. Scallan E, Hoekstra RM, Mahon BE, Jones TF, Griffin PM (2015) An assessment of the human health impact of seven leading foodborne pathogens in the United States using disability adjusted life years. Epidemiol Infect 143(13):2795–2804. doi:10.1017/S0950268814003185
8. Van den Bulcke MHG, Lievens APNR, Leunda A, MbongoloMbella EG, Barbau-Piednoir E, Sneyers MJS (2008) Transgenic plant event detection. Brussels, Belgium Patent WO 2008/092866, 2008
9. Van den Bulcke M, Lievens A, Barbau-Piednoir E, MbongoloMbella G, Roosens N, Sneyers M, Leunda Casi A (2010) A theoretical introduction to "combinatory SYBRGreen qPCR screening", a matrix-based approach for the detection of materials derived from genetically modified plants. Anal Bioanal Chem 396(6):2113–2123. doi:10.1007/s00216-009-3286-7
10. Gödel K (1931) Über formal unentscheidbare Sätze der principia Mathematica und verwandter Systeme I. Mon Math Phys 38(1):173–198. doi:10.1007/BF01700692
11. Core Team R (2015) R: a language and environment for statistical computing. R Foundation for Statistical Computing, Vienna, Austria
12. Chang W, Cheng J, Allaire JJ, Xie Y, McPherson J (2015) Shiny: web application framework for R. http://CRAN.R-project.org/package=shiny. Accessed 23 Feb 2015

Chapter 4

Detection of *Helicobacter pylori* DNA in Formalin-Fixed Paraffin-Embedded Gastric Biopsies Using Laser Microdissection and qPCR

María Fernanda Loayza Villa, Valeria Liliana Herrera Sevilla, and Nicolás Vivar-Diaz

Abstract

Molecular detection and analysis of virulence factors of *Helicobacter pylori* depends on the specificity of cell selection in the gastric biopsies. The laser microdissection (LM) instruments combine microscopy with laser cut sectioning. This combination allows one to choose only the bacteria that are in direct contact with epithelial cells in the gastric biopsy sample, avoiding those microorganisms attached to the mucus layer in the sample. The average concentration of DNA isolated from 25 cuts with selected bacteria is around 1.94 ng/μL, which is enough DNA to perform a qPCR protocol using real-time instruments to amplify 16sDNA or virulence factors like *cag*A or *vac*A. Consequently, the application of these technologies in the molecular analysis of *Helicobacter pylori* directly in contact with the surface of gastric epithelial cells is more precise and could yield better insights about the complex mechanisms of interactions between pathogen and host. Insights derived from research using the techniques described herein may in future facilitate prevention of infection or improved therapeutic options.

Key words *Helicobacter pylori*, Laser microdissection, qPCR detection, Molecular microbiome analysis, Molecular pathology

1 Introduction

The molecular detection of human pathogens is one of the major advantages of biotechnological development because of its impact on laboratory medicine [1]. These tools improve the analytical sensitivity and specificity of diagnoses and their precision decrease the time required and the cost of each test in the laboratory [1, 2]. The investment in biotechnological research and development applied to health care is worthwhile due to increased cost-effectiveness in the long term [1, 2].

The benefit of molecular techniques based on target gene amplification is indisputable, especially in the case of detecting pathogens (virus or bacteria) associated with high risk of

Kimberly A. Bishop-Lilly (ed.), *Diagnostic Bacteriology: Methods and Protocols*, Methods in Molecular Biology, vol. 1616, DOI 10.1007/978-1-4939-7037-7_4, © Springer Science+Business Media LLC 2017

pathogenesis and disease [3]. Successful tools that have been developed for diagnosis are, for example, those focused on *Mycobacterium tuberculosis* or human papilloma virus detection [3, 4]. On the other hand, when the molecular analysis is focused on *Helicobacter pylori* (*H. pylori*), it is important to consider some features of the physiopathology and the complex relation between the bacterium and the host gastric epithelial cells [5] before applying any molecular technique to detect the pathogen in gastric biopsies.

H. pylori is a microorganism that is specially adapted to the challenging environment of the gastric mucus epithelium [5–7]. There are several research reports describing *H. pylori* infection rates of around 50% of the world population, with an increased prevalence of infection in developing countries, especially in South America [8]. However, not all infected subjects show signs or symptoms of gastric disease or neoplastic development, and therefore the positive detection of *H. pylori* does not necessarily equate to illness [6–9]. There are controversial criteria about the therapy applied in consequence of *H. pylori* positive detection, because therapy could be a factor that may improve bacterial adaptation in the challenging environment of gastric mucosa and complicate the clinical condition [10–15].

Gastric pathology depends on the bacterial genotype and its ability to interact with the gastric epithelial cells through Type 4 Secretion Systems (T4SS) (Fig. 1), adhesins, and other surface proteins [15–18]. So while a given *H. pylori* genotype may cause infection and be associated with development of gastric cancer [19–20], other *H. pylori* genotypes could be part of a transition microorganism group in the mucus layer of the stomach of the host, but not really be involved in the gastric disease physiopathology [11, 13] (Fig. 2).

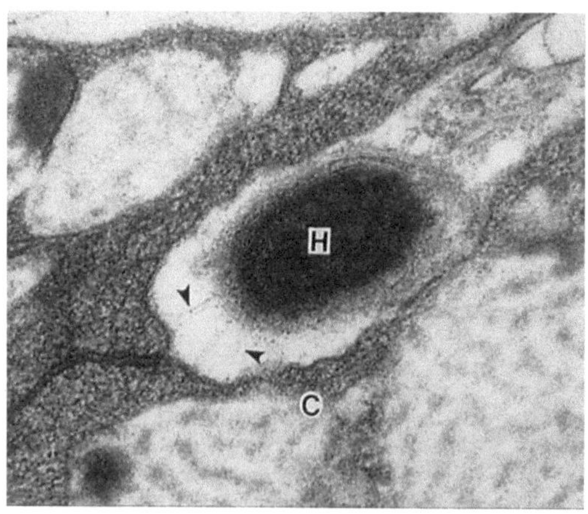

Fig. 1 Electronic microphotograph showing the *H. pylori* (**H**) interaction with epithelial cells (**C**) through Type 4 Secretion System. *Black arrows* show the projections of bacterial pili

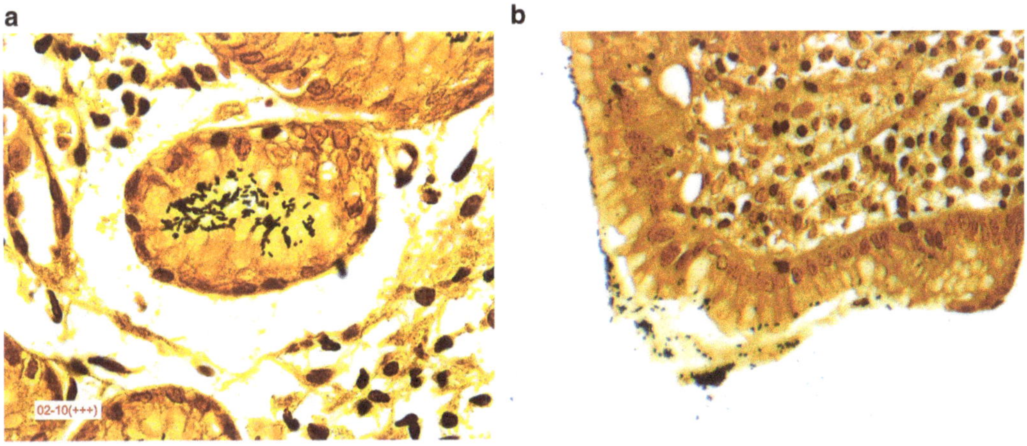

Fig. 2 Gastric tissue with Wartin Starry stain where *H. pylori* are stained with *dark black color*. (**a**) Direct interaction between bacteria (**H**) and epithelial cells (**C**) in the lumen of the gastric gland. (**b**) Bacteria (**H**) located in the mucus layer, without gastric cell (**C**) interaction

The conventional DNA isolation protocols from whole gastric biopsy samples (preserved in saline buffer or formalin-fixed, paraffin-embedded (FFPE) tissue blocks) are not able to differentiate the relationship of the bacteria with epithelial cells [20]. There are reports that show simultaneous coinfection of different genotypes in the same sample [21], but they could not differentiate which ones are probably more related to the patient condition or describe the complex molecular interactions between bacteria and host [13].

Laser microdissection (LM) consists of a microscopy-directed technique for recovery of a cluster of cells or a single cell with precision from a complex tissue section [8, 10, 22, 23]. The most common procedure for tissue analysis is paraffin embedding followed by microtome cuts of 3 μm thickness and hematoxylin–eosin (HE) staining. The processed tissue could be analyzed under microscopy and then specific tissue zones could be selected and cut for DNA isolation and molecular detection. If a gastric biopsy is analyzed under microscopy to detect those microorganisms that are in an intimate relationship with epithelial cells, with the LM, it is also possible to select them to analyze their homogeneous genetic determinants or virulence factors and associate these features with the molecular characterization of the gastric pathology. The principal objective of this protocol is to describe the procedure for isolating and amplifying genomic DNA from a homogeneous cluster of *H. pylori* in direct contact with gastric epithelial cells.

2 Materials

To improve the quality of the DNA isolated and the qPCR assays it is important to control the performance of previous steps: fixation, paraffin embedding, and staining procedures (routine histotechnical processes).

2.1 Sample Fixation and Histotechnical Process

1. 10% buffered formalin solution, pH 7.2 (*see* **Notes 1–3**).
2. One set of filter paper and pencil per biopsy sample [24].

2.2 Tissue Embedding and Processing

1. A histology cassette (41 mm × 28 mm × 6 mm) with ID number for each patient.
2. For paraffin embedding [24]:
   (a) 10% buffered formalin.
   (b) 75% ethanol.
   (c) 80% ethanol.
   (d) 90% ethanol.
   (e) Absolute ethanol.
   (f) Clearing solution (*see* **Note 4**).
   (g) 4.3 L melted paraffin (56–58 °C).
3. Enclosed tissue processor.
4. Wing hand warm tweezers, spatula, and metal blocks.
5. Ice bath.
6. One charged glass slide and a nuclease free slide coated with polyethylene naphthalate (PEN) membrane (*see* **Note 5**).
7. Semiautomatic microtome and new coated microtome blades.
8. Flotation water bath, set at 45 °C.
9. Tissue baking processor set at 45 °C (*see* **Note 6**).
10. Auto stainer instrument with ready-to-use reagent system [25]:
    (a) Clearing solution.
    (b) Absolute ethanol.
    (c) Ethanol 96%.
    (d) Ethanol 70%.
    (e) Eosin.
    (f) 0.2% ammonia water.
    (g) Hematoxylin.
    (h) Demineralized water.
11. Microscope cover glass (24 mm × 60 mm × 0.21 mm).
12. Mounting medium.

2.3 Selection and Cut of Bacteria in Tissue Sections

1. Laser microdissection system.
2. Cap microcentrifuge tubes adapted to LM system (*see* **Note 7**).
3. Alcohol-resistant pens to identify the tubes.

2.4 DNA Isolation and Amplification by qPCR

The DNA isolation process from samples caught in the adhesive cap tube has to be performed with high-quality reagent sets. It is important to use columns to isolate genomic DNA. To perform the isolation and amplification, it is necessary to prepare.

1. BSC IIA cabinet.
2. Two dry heating blocks, one heating block set at 56 °C and the other heating block set at 70 °C.
3. Microcentrifuge and microcentrifuge rotor for 2.0 mL tubes (Eppendorf 5415D BenchTop Microcentrifuge).
4. Vortex.
5. 1.5 or 2.0 mL microcentrifuge tubes for wash and elution steps.
6. Pipette tips with filter to avoid cross-contamination.
7. Micro pipettes 1–10 μL, 10–100 μL, 100–1000 μL.
8. Disposable gloves.
9. QIAamp DNA Micro Kit (Qiagen) equilibrated to room temperature
   (a) Molecular-grade ethanol.
   (b) AW1 and AW2 buffers prepared following the manufacturer instructions.
   (c) RNA carrier solution prepared by mixing 50 μL of AL buffer and 1 μL of RNA carrier for each sample.
10. Picogreen reagent set (INVITROGEN™), equilibrated to room temperature.
11. Fluorometer, turned on 30 min before the plate reading.
12. qPCR primers for *H. pylori* detection and initial molecular analysis (*see* **Note 8**).
13. 0.2 mL PCR tubes (eight tubes in chain).
14. SsoAdvanced™ universal qPCR supermix with SYBR® Green, prepared as per manufacturer's instructions.
15. Real-time PCR instrument (such as Bio-Rad CFX96 Real Time System C1000 Touch Thermal Cycler), turned on 30 minutes before assay.

3 Methods

In order to guarantee the quality of the procedure, it is important to consider the following steps: (1) sample fixation and histotechnical process; (2) tissue embedding and processing; (3) selection and cut of bacteria in tissue; (4) DNA isolation and amplification by qPCR [24].

3.1 Sample Fixation and Histotechnical Process

The physician takes five biopsy samples from different stomach zones during the endoscopy procedure in accordance with the Sydney system (*see* **Note 9**) [26]. It is important to consider that the major aim of the biopsy is to detect inflammation, neoplastic changes and *H. pylori* infection [27]. The stomach zone of each biopsy sample has to be identified with pencil and placed in filter paper to avoid unintentional erasure of identifying labels from each sample during the following steps [24–26].

Typically, each biopsy is around 3 mm thickness [24]. Immediately after the samples have been taken from the stomach, they need to be fixed using 10% buffered formalin, pH 7.2 (Table 1). Due to the size of biopsies, the samples have to be processed within 5 h after they were fixed (*see* **Note 3**) [24].

3.2 Tissue Embedding and Processing

1. Place the fixed and paper involved samples of each patient in a histotechnique plastic cassette.
2. Program the processing schedule (Table 2) in the instrument software (*see* **Note 10**).
3. The processing protocol consists of the consecutive changes of hydrophilic to hydrophobic solutions to change the water inside the tissue matrix for paraffin (*see* **Note 11**). It is important to control the replacement of solutions (listed in Table 2) for an optimal processing [24]. It is possible to process the tissue manually, but it is time intensive and may decrease the quality of the results. The tissue has to be completely submerged in each solution during the procedure. Generally, the entire processing takes around 18 h. The tissue cassettes may remain in melted paraffin overnight [24, 25].
4. Prepare the embedding workstation instrument it is necessary to turn on the instrument 1 h before start to work. Complete the paraffin reservoir if it is necessary.
5. To prepare paraffin blocks of tissue, it is important to orient the samples from the antrum to the incisura; this orientation

Table 1
Formula for 10% buffered formalin preparation (pH 7.2 must be controlled in each prepared batch)

Reagent	Required volume (mL)
Formol 37–40%	100
PBS 10×	36
Deionized water	864

Table 2
General processing schedule for paraffin embedding of tissue from formalin-fixed biopsies

Step	Reagent	Function	Immersion time (h)
1	10% buffered formalin	Fixative	2
2	75% ethanol	Dehydrant	1
3	80% ethanol	Dehydrant	1
4	90% ethanol	Dehydrant	1
5	Absolute ethanol	Dehydrant	1
6	Absolute ethanol	Dehydrant	1
7	Absolute ethanol	Dehydrant	1
8	Clearing solution	Clearing agent	2
9	Clearing solution	Clearing agent	2
10	Melted paraffin (60 °C)	Embedding agent	2
11	Melted paraffin (60 °C)	Embedding agent	2

facilitates the observation of pathology and subsequent diagnosis (*see* **Note 12**) [26]. With warm tweezers, attach the biopsies at the bottom of the metal block; it may be useful to cool the base of the block slightly in the cold plate (Fig. 3a).

6. Add melted paraffin to cover the tissue and complete the block volume. The same tissue cassette must be used to support and identify the block.
7. Place the block in the cold plate and wait until the paraffin is completely solidified. Typically, this takes at least 1 h.
8. Demold the paraffin block using a spatula and place the blocks on ice (*see* **Note 13**) (Fig. 3b).
9. Subsequently, the paraffin blocks have to be smoothed with a semiautomatic microtome.
10. Cut sections of 3 μm thickness from the tissue block with a new disposable blade.
11. Place the block tape on a tissue flotation water bath and stretch it with a fine tweezers. Be sure that the tissue contained on the tape does not have wrinkles (*see* **Note 14**).
12. Catch one section of the tape on a positive charged glass slide and other one on the membrane slide (*see* **Note 15**).
13. Place the positive charged glass slide on the tissue baking processor to remove excess paraffin. This step is not necessary

Fig. 3 Paraffin embedding and processing of gastric tissue samples. (**a**) It is important to consider the correct orientation of the tissue in the block. With a warm tweezers, ensure the orientation of the tissue in the bottom of metallic block. (**b**) Let the paraffin block solidify in ice bath before microtome cutting to obtain tapes of 3 μm thickness

if the dewax process can be performed in an auto stainer instrument.

14. Place both, the positive charged glass slide and membrane slide in the auto stainer instrument to perform the HE staining protocol. The expected staining results are: nuclei stained in blue and cytoplasm pink to red. The conventional staining program consists of two steps: (a) Dewaxing phase which is described in Table 3 and (b) staining protocol (Table 4). If an auto stainer is not available, it is possible to perform a handmade staining procedure in Coplin jars (*see***Note 16** and **17**). The auto stainer instrument must be set so that the membrane slide is not mounted. Alternatively, a Wartin Starry stain (Table 5) could be performed if only *H. pylori* observation is important in the analysis of the tissue. This special stain protocol may be performed with the positive charged glass slide (*see***Note 18**). The Wartin Starry staining consists of the following steps:
    (a) Dewax a tissue cut of 3 mm of thickness and hydrate with working glycine–acetic acid solution.
    (b) Place the glass slides in 40 mL of 0.25% silver nitrate in a Coplin jar, with the cap. Incubate the slides at 75–80 °C for 2 min. The solution could be heated using a microwave.
    (c) Place the glass slides in freshly prepared silver nitrate–gelatin–hydroquinone developer in a Coplin jar, with cap, heat the solution in microwave (power level 1 for 1 min).

Table 3
List of solutions for dewaxing tissue cuts in glass slides or membrane slides

Step	Reagent	Function	Immersion time
1	Xylene or Histoclear	Dewax	5 min
2	Xylene or Histoclear	Dewax	5 min
3	Absolute ethanol	Rehydrate	20 s
4	Absolute ethanol	Rehydrate	20 s
5	90% ethanol	Rehydrate	20 s
6	70% ethanol	Rehydrate	20 s
7	Tap water	Rinse	

Table 4
Hematoxylin–eosin in-house conventional staining protocol in Coplin jars (without auto stainer instruments)

Step	Reagent	Function	Immersion time
1	Tap water	Wash	
2	Hematoxylin solution	Dye	1 min
3	Tap water	Wash	6 slide immersions
4	Eosin Y solution	Contrast stain	8 slide immersions
5	95% ethanol	Wash	8 slide immersions
6	95% ethanol	Dehydrate	8 slide immersions
7	Absolute ethanol	Dehydrate	8 slide immersions
8	Absolute ethanol	Dehydrate	8 slide immersions
6	Xylene	Final wash	4 slide immersions
7	Xylene	Mounting solvent	4 slide immersions

Agitate the Coplin jar for about 15 s. Incubate the slides 1 min or until the sections appear gray.

(d) Wash quickly and thoroughly in hot running water.

(e) Rinse in two changes of distilled water, counterstain with tartrazine solution for 15 s

(f) Rinse in two changes of distilled water.

(g) Dehydrate, clear, and mount as a routine stained slide.

Table 5
Wartin Starry solution preparation and manual staining performance [28]

Glycine–Acetic Acid 100× (Stock solution)	
Glycine	1.2 g
Acetic acid, glacial	0.3 mL
Distilled water	100.0 mL
pH around 3.5	
0.25% Silver Nitrate Solution	
Silver nitrate	0.1 g
Glycine–acetic acid working solution (1×)	40.0 mL
2% Silver Nitrate Working Solution	
Silver nitrate solution	0.2 mL
Glycine–acetic acid working solution (1×)	10.0 mL
0.2% Hydroquinone Solution	
Hydroquinone 0.2 g	0.2 g
Glycine–Acetic Acid Working solution (1×)	10.0 mL
4% Gelatin Solution	
Gelatin	1 g
Glycine–acetic acid working solution (1×)	25.0 mL
This solution has to be dissolved on hot magnetic plate	
0.2% Hydroquinone Solution	
Hydroquinone	0.02 g
Glycine–acetic acid working solution (1×)	10.0 mL
Place the 2% silver nitrate, 4% gelatin, and the 0.2% hydroquinone solutions in separate flasks in a 55–58 °C oven. These solutions and the 0.25% silver nitrate are prepared immediately before use	
Silver Nitrate–Gelatin–Hydroquinone Developer	
Silver nitrate, 2%	10.0 mL
Gelatin, 4%	25.0 mL
Hydroquinone, 0.2%	10,0 mL
Mix the solutions in the order described. Ensure the solutions are mixed well after each addition. Prepare immediately before use	
0.1% Tartrazine Solution	
Tartrazine, C.I. 19,140	0.1 g
Distilled water	100.0 mL
Acetic acid	0.2 mL

15. Use mounting medium to mount the membrane slide. The cover slide must be placed in the back side of the membrane slide covering the stained tissue (Fig. 4), but the mounting medium must be only in the border of the slide. Do not put the mounting medium over the PEN membrane.
16. Place the cover slide on the back side of the membrane slide and wait until it is completely dry.

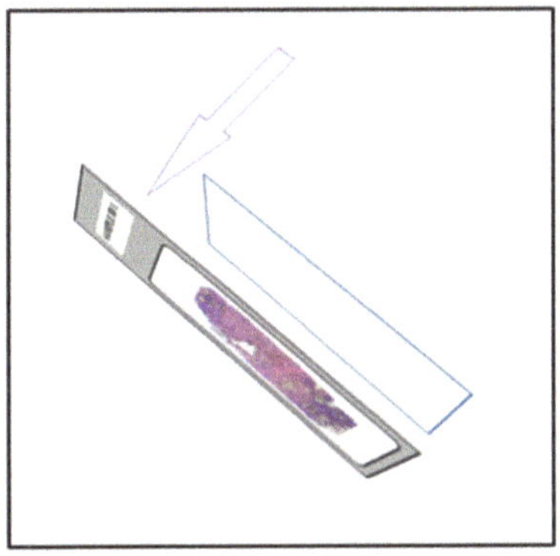

Fig. 4 The cover slide must to be mounted on the back side of the membrane slide in contact with stained tissue, but mounting medium must not be added over the PEN membrane

3.3 Selection and Tissue Sectioning

One tissue tape collected on PEN membrane slide will be used in the selection and sectioning of samples, to be used for molecular analysis of *H. pylori* attached to the gastric epithelial cells.

To examine the exact localization of the bacteria in the gastric mucosa, use the positive charged glass slide (*see* **Note 19**) [23].

1. Follow the operator's manual of LM system to set the slide location with the software options (*see* **Note 20**).
2. Set the power, velocity, and thickness of the laser according to the operator's manual instructions (*see* **Note 21**).
3. Place the membrane slide on the stage. Be sure that the membrane with tissue side is down, because the laser system cuts from below and the cut section is captured in the adhesive lid from above.
4. Select the cut sections. First, it is important to compare the images of the tissue between glass slide and membrane slide. The sections to be selected are those with ten or more bacteria directly located on the surface of epithelial cells in the lumen of the gastric glands. Do not select bacteria included in the mucus layer in the sample.
5. With software options draw a figure to enclose the tissue section to be cut and captured. Avoid selecting the epithelial cell nucleus. Each selection could be around 10–20 μm of diameter. Collect at least 25 cut sections with the same conditions in the tube lid for DNA isolation.

3.4 DNA Isolation and Amplification by qPCR

1. For genomic DNA isolation it is important to equilibrate all components of the QIAamp DNA Micro Kit to room temperature. Alternative DNA isolation kits could be used provided that quality DNA is isolated (column based kits are preferable).

2. Since a small sample is available and the cut sections are adhered in the tube lid, the first step of the extraction has to be performed with 30 μL of Buffer ATL and 20 μL proteinase K. Add these reagents into the 0.2 mL tube containing each laser microdissected sample.
3. Mix the solution and sample by vortexing for 15 s and then place the tubes in a heat block at 56 °C with the lid down for 6 h to lyse the tissue sticking to the tube cap. Pulse-vortex occasionally.
4. Examine the lid with a magnifying glass to ensure that all of the sample section is lysed.
5. Add 50 μL of ATL to each sample and vortex to mix.
6. Transfer each lysate sample to a new 1.5 mL microcentrifuge tube.
7. Add 50 μL of working AL solution (1 μL of RNA carrier in 50 μL Buffer AL for each sample), following manufacturer's instructions.
8. Add 100 μL molecular-grade ethanol and vortex for 15 s to mix. Incubate the sample for 5 min at room temperature.
9. Centrifuge for 10 s to spin down drops.
10. Transfer 200 μL of the lysate sample to the QIAamp MinElute column placed into a 2 mL collection tube. Centrifuge the sample at 6,000 × *g* for 1 min. Place the column in a new collection tube.
11. Wash the column with 700 μL Buffer AW1; centrifuge at 6,000 × *g* for 1 min. Place the column in a new collection tube and repeat this washing step with 700 μL of AW2 buffer.
12. Perform an additional centrifuge step at 20,000 × *g* for 3 min to remove excess washing solutions and dry the membrane.
13. Replace the collection tube with a 1.5 mL sterile microcentrifuge tube labeled with sample identification.
14. Add 35 μL of elution buffer (AE solution) dispensed in the center of the column. Incubate for 5 min at 20–25 °C and centrifuge at 20,000 × *g* for 1 min. The isolated DNA can be stored at −20 °C until it is used in a qPCR assay.
15. To evaluate the concentration of genomic DNA isolated, the fluorometric method is recommended using Picogreen, (Invitrogen™) as label dye, following standard instructions recommended by the manufacturer. This method is suggested because it is important to evaluate the quality of DNA isolated. NanoDrop technology only measures the concentration of nucleotides in the solution, but it does not mean that the sample contains an intact DNA molecule. On average, concentration of DNA isolated from 25 laser microdissection cut sections is around 1.96 ng/μL ± 0.16 ng/μL [23].

Amplification of 16s rDNA of *H. pylori* from isolated genomic DNA may be used to confirm the detection of the bacteria, but in this protocol it is used as internal control. If there is a good amplification curve, it means that conserved DNA molecule has been isolated. Other genes can be evaluated using the same protocols of sample collection. The amplification products of *cag* and *vac* genes, for example, are described in Table 6 (Figs. 5 and 6). The reagent cocktails for amplification and the qPCR program are described in Tables 7 and 8, respectively.

The technical equipment and instruments described in this writing paper are only a recommendation, and they are named because of their availability and utility in the laboratory. There are alternative commercial instruments that can be useful for the same application described here.

4 Notes

1. Prepare 10% buffered formalin solution with a neutral pH to preserve the chemical and structural constitution of the tissue by inhibition of proteolytic activity and environmental bacteria contamination. The 10% buffered formalin solution preparation is described in Table 1.

Table 6
Recommended primers for detection of *H. pylori* with 16s rDNA and initial molecular analysis for virulence factors like *cag*A and *vac*A for qPCR performed in BIO-RAD CFX96 Real Time System (C1000 Touch Thermal Cycler)

Primer set	Description	Product size	Melting temperature
16s rDNA [29]	>16S FORWARD TGCGAAGTGGAGCCAATCTT >16S REVERSE GGAACGTATTCACCGCAACA	118 bp	82.5 °C
*cag*A [29]	>CAG-A-FORWARD ATAATGCTAAATTAGACAACTTGAGCGA >CAG-A-REVERSE TTAGAATAATCAACAAACATCACGCCAT	298 bp	78 °C
*vac*A s1/s2 [30]	>VacAs1-s2 FOWARD ATGGAAATCAACAAACACAC >VacAs1-s2 REVERSE CTGCTTGAATGCGCCAAAC	*vac*A s1 259 bp *vac*A S2 286 bp	 83.5 °C 84.5 °C
*vac*A m1/m2 [30]	>VacAm1-m2 FOWARD CAATCTGTCCAATCAAGCGAC >VacAm1-m2 REVERSE GCGTCAAAATAATTCCAAGG	*vac*A m1 567 bp *vac*A m2 642pb	 81.5 °C 82 °C

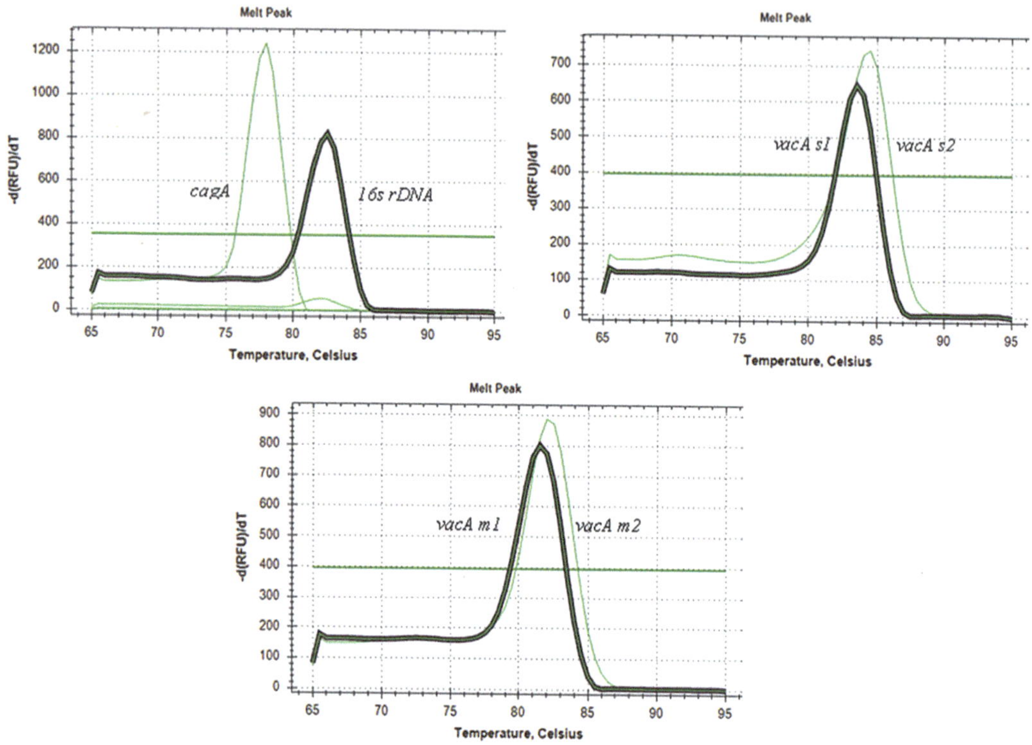

Fig. 5 Melting curves for qPCR products obtained from amplifications of *H. pylori* 16s rDNA, *cagA*, and *vacA* genes from DNA isolated from laser microdissection zones selected from gastric biopsy

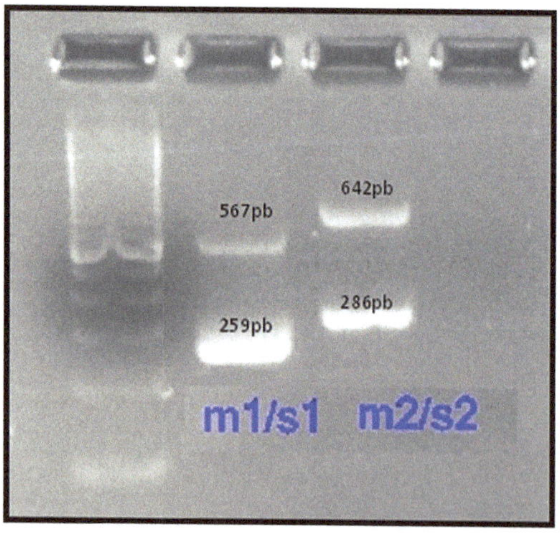

Fig. 6 Agarose gel electrophoresis showing the product size of *vac* A amplification by qPCR, using DNA isolated from laser microdissected sections of gastric biopsy

Table 7
Amplification cocktail reagents prepared for qPCR amplification using SsoFast™ EvaGreen® Supermix

Reagents	Standard concentration	Used volume (15 μL)
SsoFast™ EvaGreen® Supermix (Bio-Rad)	2×	7.5 μL
Forward primer	10 μM	1 μL
Reverse primer	10 μM	1 μL
Molecular biology grade water	–	0.5 μL

Table 8
Amplification programs performed in Bio-Rad CFX96 Real Time System (C1000 Touch Thermal Cycler)

	16sDNA		*cagA*		*vacA m1/m2* y *vacA s1/s2*	
Initial denaturation	95 °C	10′	95 °C	10′	95 °C	10′
Denaturation	95 °C	10″	95 °C	10″	95 °C	1′
Annealing	60.7 °C	20″	61.5 °C	20″	56.7 °C	1′
Elongation	72 °C	15″	72 °C	20″	72 °C	1′
Cycle number	35		35		40	

2. Control the pH of each preparation and maintain the volume relation sample–formalin (10:1) [24]. If the formalin solution is acid, dark precipitates may be observed in blood vessels in the tissue analysis; these precipitates correspond to acid hematin formed by the reaction between hemoglobin and acid formalin buffer. Also, some cellular structures may not be stained appropriately.

3. Some important details in fixation are temperature, volume of fixative solution used, size of tissue sample, and time of fixation. If there is an excess of fixative solution or fixation time, formalin could react with molecules in the surface of the cells and also destroy DNA molecules. Consequently, the sample could not be used in immunohistochemistry or molecular analysis.

4. Xylene is an organic solvent that was used as a clearing and deparaffinizing solution during tissue processing and staining; however, it is a toxic solution with several side effects. Nowadays, there are some commercial xylene substitutes

available, for example: Histo-Clear, Neoclear®, Clearify™, MasterClear™, Naturalene™. All of these solutions can be used as clearing agents.

5. Nuclease0free slides are used for molecular analysis of the tissue cuts obtained from LM. There are several commercial alternatives of the nuclease free membrane slides.
6. It is not necessary if an automated stainer system is available.
7. These nuclease-free microcentrifuge tubes are specific for each LM system, and there are different tube's volumes. We recommend 0.5 mL tubes for latter molecular analysis.
8. Primers for detection of *H. pylori* with 16s rDNA and virulence factors like *cagA* and *vacA* using qPCR are just a suggestion (Table 6). Any set of primers can be used for detection of virulence or resistance genes associated with *H. pylori.*
9. The Sydney system consists of five biopsy sites: two antral biopsies, two of gastric corpus, and one of incisura.
10. The processing schedule can be changed in accordance with the available instruments in the laboratory; it is necessary to standardize the process in each laboratory.
11. If the instrument is not available, prepare the solutions needed to replace the paraffin included in the tissue for hydrophilic solutions listed in Table 2.
12. Before the inclusion, pay attention to the orientation of the tissue; it is important to place the tissue over one side, targeting the mucosal layer up and the muscular layer of the gastric mucosa down.
13. Maintain the paraffin block on ice for better cutting in the microtome.
14. When molecular analysis is to be performed on the tape, it is important to change the water bath for each sample to avoid cross-contamination.
15. The tissue tape has to be collected on the back side of the membrane slide; this location protects the tissue until the microdissected section will be cut and selected in the cap of the microcentrifuge tube.
16. To perform a nonautomatated conventional stain process with hematoxylin–eosin it is necessary to have Coplin jars, the slides need to be submerged in the staining solutions like it is described in Table 4. The formula for the preparation of each dye component of the stain procedure is described in Table 9.
17. It is better if preprepared Harris hematoxylin is acquired commercially. If it is prepared in your laboratory, avoid the use of mercury oxide.

Table 9
Hematoxylin–eosin traditional formula for preparation of dye solutions

Hematoxylin solution			**Eosin solution**		
Reagent		**Function**	**Reagent**		**Function**
Hematoxylin	5.0 g	Dye	Ethanol	800 mL	Solvent
Ammonium alum	50 g	Mordant	Distilled water	200 mL	Solvent
Distilled water	700 mL	Solvent	Glacial acetic acid	5 mL	Acidifier
Absolute ethanol	50 mL	Solvent	Eosin Y ws	2 g	Dye
Calcium hypochlorite	40 g	Oxidant			
Glacial acetic acid	50 mL	Acidifier			

18. Wartin Starry stain in the duplicate sample glass slide can show the *H. pylori* infection, because it stains the bacteria in black color with high specificity. All microdissected sections are selected on the basis of initial microscopic evaluation and only *H. pylori* positive samples are included in the analysis. The molecular analysis can be performed to investigate the presence of antibiotic resistance genes, to classify the *H. pylori* genotype or to analyze the specific expression of virulence factors. 16s rDNA has to be used as a qPCR internal control of amplification when laser microdissection is applied.
19. The resolution of tissue images in the microscope adapted to the LM system is poor because the air between sample and the cover glass affect the refractive index. Therefore, it is important to have a glass slide, stained with hematoxylin–eosin or Wartin Starry, which is perfectly mounted, to analyze the exact location in the tissue that has to be cut with a laser.
20. The laser microdissection system's operator has to be trained before using the instrument.
21. The power, velocity, and thickness of the laser should be calibrated with each sample's batch.

References

1. Versalovic J, Lupski JR (2002) Molecular detection and genotyping of pathogens: more accurate and rapid answers. Trends Microbiol 10:15–21. doi:10.1016/S0966-842X(02)02438-1
2. Bibby K (2013) Metagenomic identification of viral pathogens. Trends Biotechnol 31:275–279. doi:10.1016/j.tibtech.2013.01.016
3. Abreu ALP, Souza RP, Gimenes F, Consolaro MEL (2012) A review of methods for detect human Papillomavirus infection. Virol J 9:262. doi:10.1186/1743-422X-9-262
4. Nahid P, Pai M, Hopewell PC (2006) Advances in the diagnosis and treatment of tuberculosis. Proc Am Thorac Soc 3:103–110. doi:10.1513/pats.200511-119JH
5. Alzahrani S (2014) Effect of Helicobacter pylori on gastric epithelial cells. World J Gastroenterol 20:12767. doi:10.3748/wjg.v20.i36.12767

6. Calvet X, Ramírez Lázaro M-J, Lehours P, Mégraud F (2013) Diagnosis and epidemiology of *Helicobacter pylori* infection. Helicobacter 18:5–11. doi:10.1111/hel.12071
7. Choi YJ, Kim N, Lim J et al (2012) Accuracy of diagnostic tests for Helicobacter pylori in patients with peptic ulcer bleeding. Helicobacter 17: 77–85.doi:10.1111/j.1523-5378.2011.00915.x
8. de Bernard M, Josenhans C (2014) Pathogenesis of *Helicobacter pylori* infection. Helicobacter 19:11–18. doi:10.1111/hel.12160
9. Eusebi LH, Zagari RM, Bazzoli F (2014) Epidemiology of Helicobacter pylori infection. Helicobacter 19:1–5. doi:10.1111/hel.12165
10. Testerman TL (2014) Beyond the stomach: an updated view of *Helicobacter pylori* pathogenesis, diagnosis, and treatment. World J Gastroenterol 20:12781. doi:10.3748/wjg.v20.i36.12781
11. Salama NR, Hartung ML, Müller A (2013) Life in the human stomach : persistence strategies of the bacterial pathogen *Helicobacter pylori*. Nat Rev Microbiol 11(6):385–399. doi:10.1038/nrmicro3016
12. Plummer M, Franceschi S, Vignat J et al (2014) Global burden of gastric cancer attributable to H. pylori. Int J Cancer 136:487–490. doi:10.1002/ijc.28999
13. Abreu MT, Peek RM (2014) Gastrointestinal malignancy and the microbiome. Gastroenterology 146:1534–1546.e3. doi:10.1053/j.gastro.2014.01.001
14. Micu G, Stăniceanu F, Zurac S et al (2010) Carcinogenesis and infection with Helicobacter pylori. Rom J Intern Med 48:299–306
15. Backert S, Selbach M (2008) Role of type IV secretion in Helicobacter pylori pathogenesis. Cell Microbiol 10:1573–1581. doi:10.1111/j.1462-5822.2008.01156.x
16. Saxena A, Shukla S, Prasad KN, Ghoshal UC (2011) Virulence attributes of Helicobacter pylori isolates & their association with gastroduodenal disease. Indian J Med Res 133:514–520
17. Yamaoka Y (2010) Mechanisms of disease: Helicobacter pylori virulence factors. Nat Rev Gastroenterol Hepatol 7:629–641. doi:10.1038/nrgastro.2010.154
18. Loh JT, Shaffer CL, Piazuelo MB et al (2011) Analysis of cagA in Helicobacter pylori strains from Colombian populations with contrasting gastric cancer risk reveals a biomarker for disease severity. Cancer Epidemiol Biomark Prev 20:2237–2249. doi:10.1158/1055-9965.EPI-11-0548
19. Konturek PC, Konturek SJ, Brzozowski T (2009) Helicobacter pylori infection in gastric cancerogenesis. J Physiol Pharmacol 60:3–21
20. Correa P, Piazuelo MB (2012) Evolutionary history of the Helicobacter pylori genome: implications for gastric carcinogenesis. Gut Liver 6:21–28. doi:10.5009/gnl.2012.6.1.21
21. Ben Mansour K, Fendri C, Zribi M et al (2010) Prevalence of Helicobacter pylori vacA, cagA, iceA and oipA genotypes in Tunisian patients. Ann Clin Microbiol Antimicrob 9:10. doi:10.1186/1476-0711-9-10
22. Simone, N. L., Bonner, R. F., Gillespie, J. W., Emmert-Buck, M. R., & Liotta, L. a. (1998). Laser-capture microdissection: Opening the microscopic frontier to molecular analysis. Trends in Genetics, 14(7), 272–276. https://doi.org/10.1016/S0168-9525(98)01489-9
23. Loayza MF, Villavicencio FX, Santander SC et al (2015) Improved method for extraction and detection of Helicobacter pylori DNA in formalin-fixed paraffin embedded gastric biopsies using laser micro-dissection. MethodsX 2:1–7. doi:10.1016/j.mex.2014.11.003
24. Vivar N (2010) Manual de procedimientos en anatomía patológica, 1st edn. ROCHE, Quito-Ecuador
25. Rolls G (2016) Performing a hematoxylin and eosin stain (H&E). Leica Science Lab Tutorial. http://www.leicabiosystems.com/pathologyleaders/performing-a-hematoxylin-and-eosin-stain-a-step-by-step-guide. Accessed 29 Dec 2015
26. De la Torre Bravo A (2009) Procedimientos endoscópicos en gastroenterología, 2nd edn. Médica Panamericana. Córdova Villalobos, D.T.B.
27. Lee JY, Kim N (2015) Diagnosis of Helicobacter pylori by invasive test: histology. Ann Transl Med 3:10. doi:10.3978/j.issn.2305-5839.2014.11.03
28. Churukian C (2009) Microwave modification of the Warthin Starry method for bacteria, 2nd edn. University of Rochester Medical Center. http://www.urmc.rochester.edu/urmc-labs/pathology/stainsmanual/index.html?MICROWAVEMODIFICATIONOFTHEWARTHIN-STARRYMETHODFORBACTERIA. Accessed 20 Dec 2015
29. Sepúlveda E, Moreno J, Spencer ML, Quilodrán S, Brethauer U, Briceño C, García A (2012) Comparación de *Helicobacter pylori* en la cavidad oral y mucosa gástrica de acuerdo a genotipo de virulencia (cagA y vacAm1). Rev Chil Infectol 29(3):278–283. doi:10.4067/S0716-10182012000300005
30. Herrera V (2015) Genotipificación de *Helicobacter pylori* a traves de PCR en Tiempo real a partir de biopsias gástricas parafinadasde pacientes de Esmeraldas y Quito. Disertation. Universidad de las Fuerzas Armadas ESPE

Chapter 5

Mycobacterial Load Assay

Stephen H. Gillespie, Wilber Sabiiti, and Katarina Oravcova

Abstract

Tuberculosis is a difficult disease to treat, a process made more harder as tools to monitor treatment response only provide a result long after the patient has provided a sample. The mycobacterial load assay (MBLA) provides a simple molecular test to quantify and determine the viability of *M. tuberculosis* in human or other samples.

Key words Mycobacterial load assay (MBLA), Sputum, Ribosomal RNA (rRNA), Reverse transcriptase real-time quantitative polymerase chain reaction (RT-qPCR)

1 Introduction

The mycobacterial load assay (MBLA) is a culture-free biomarker that aims at providing rapid quantification of viable *Mycobacterium tuberculosis* in patient sputum. The assay monitors the molecular load of *M. tuberculosis* (Mtb) cells in sputum samples providing accurate information on bacterial response (decline) to antimicrobial treatment. The bacterial load is determined by detection and quantification of Mtb 16S ribosomal RNA (16S rRNA) using reverse transcriptase real-time quantitative polymerase chain reaction (RT-qPCR). The 16S rRNA is a structural RNA occurring in multiple copies in a bacterial cell and its half-life and stability is considerably greater in comparison to mRNA. This makes 16S rRNA a suitable marker for the detection and quantification of viable bacterial load as it is associated with intact viable cells. When Mtb cells are killed by anti-TB therapy, the amount of 16S rRNA also declines making it possible to estimate the number of viable cells remaining in patient sputum. This quantification correlates with viable detection by solid and liquid culture (Mycobacterium Growth Indicator Tube, MGIT) [1, 2]. MBLA directly quantifies bacterial load in sputum while MGIT gives an indirect measurement based on time to detection (time taken for the sample to flag positive). The higher the bacterial load in patient sputum the shorter the time to

Kimberly A. Bishop-Lilly (ed.), *Diagnostic Bacteriology: Methods and Protocols*, Methods in Molecular Biology, vol. 1616, DOI 10.1007/978-1-4939-7037-7_5, © Springer Science+Business Media LLC 2017

positivity (TTP) on MGIT. In comparison to MGIT, MBLA is faster, insensitive to contamination and has the potential to detect nonculturable bacilli [3, 4] MBLA overcomes these inconsistencies in bacterial counts as the assay targets 16S rRNA, which is independent of the culturability of the sputum sample.

MBLA is a two-step process consisting of: (1) total RNA extraction and enzymatic removal of genomic DNA, (2) RT-qPCR and data analysis where cycle threshold (C_T) is converted to bacterial load. A cell-based extraction (internal) control is incorporated into the test samples prior to extraction in order to monitor for extraction efficiency and downstream PCR performance. A set of Mtb RNA preparations with known concentrations is amplified along with the test samples to provide a standard curve, a reference for translating C_T values into bacterial load (CFU/mL) in original sputum samples.

Using this methodology it is possible to follow the decline in bacterial load of patients being treated for tuberculosis. Since this is a molecular method, the results are available much more rapidly (within hours) than other methods dependent on culture. More importantly, the specificity of the reaction means that a false positive result from contamination is rare whereas this is a considerable cause of data loss for culture-based methods.

2 Materials

2.1 Instruments

1. Laboratory scales.
2. Fume hood.
3. Fridge (0–8 °C).
4. Freezer (−20 °C).
5. Freezer (−80 °C).
6. Vortex, e.g., Genie2.
7. Benchtop centrifuge (with sealed removable buckets, for 15 mL plastic tubes), for speeds of 3000 × *g* for 30 min, e.g., Megafuge 16, Heraeus.
8. Refrigerated microcentrifuge (with sealed removable rotor, for 1.5 mL microcentrifuge tubes and 2 mL homogenization tubes), e.g., Fresco 21, Heraeus.
9. Homogenizer, e.g., Fastprep, MP Biomedical (setting P60, 40 s) or Precellys 24, PeqLab 5000 × *g* 40 s).
10. Thermomixer, with thermoblock for 1.5 mL microtubes, e.g., BLD-455-010C, microcentrifuge tube.
11. QiAgility Pipetting Robot and accessories, e.g., QIAgility software, Qiagen plates (5-well master mix plate, 16-well reagent plate, 32-well sample plate, 72-well and 96-well reaction plates).
12. Real-time PCR thermocycler with accessories, e.g., 36-well 72-well and 100-well rotors for RotorGene Q.

2.2 Additional Equipment, Consumables, and Plasticware

1. Pipettes and matching sterile filtered pipette tips, DNAse and RNA-ase-free, range: P1000, P200, P10, P2.
2. Sterile Pasteur pipettes, 1.5 mL, 3 mL.
3. Disposable gloves, chemical resistant.
4. Safety goggles, chemical resistant.
5. Biological waste discard jars.
6. Chemical waste discard jars.
7. 500 mL plastic containers, e.g., 734-5087, Nalgene.
8. Measuring cylinders, plastic.
9. Racks for 1.5 mL and 2 mL microtubes and for 15 mL and 50 mL Falcon tubes, chemical-resistant and autoclavable.
10. Sterile disposable universal tubes and 15 mL and 50 mL Falcon tubes.
11. Sterile RNA-ase-free microtubes (1.5 mL microcentrifuge tubes), suitable for storing at −80 °C.
12. Homogenization tubes, 2 mL with screw caps, compatible with the homogenizer and microcentrifuge; filled with homogenization beads (0.1 mm glass beads), e.g., MP Biomedicals or 91-PCS-VK01, PeqLab.
13. Cooling rack for 0.5, 1.5, and 2 mL microtubes or ice bucket.
14. PCR reaction tubes suitable for the PCR instrument and the total number of samples to be analyzed. For RotorGene Q (Qiagen) use:
    - Single 0.2 mL PCR optical thin wall flat cap microtubes (any supplier), for a 32-well rotor.
    - Strips of four tubes and caps, 0.1 mL (Qiagen), for a 72-well rotor.
    - Rotor-Disc 100 with heat-seal film (Qiagen), for a 100-well rotor.

2.3 Reagents

1. Absolute ethanol (99–100%).
2. TB disinfectant, e.g., Tristel Fuse (freshly prepared or prepared within a week).
3. Molecular grade water, DNase and RNA-ase free, e.g., 13138533, Fisher Scientific.
4. Solution to remove RNases from work surfaces and equipment, e.g., RNA-ase Away (Removes RNA-ase enzymes from working space), e.g., 10666421, Fisher Scientific.
5. Guanidine thiocyanate (GTC) working solution (200 g GTC., 40 mL 1 M Tris–HCl pH 7.5, 4 mL β-mercaptoethanol, 120 mL molecular grade water (Table 1 and *see* **Note 1**).
6. 1 M Tris–HCl pH 7.5.

Table 1

Preparation of guanidine thiocyanate solution

Reagent	Amounts for 400 mL
Guanidine thiocyanate (GTC)	200 g
1 M Tris–HCl pH 7.5	40 mL
β-mercaptoethanol	4 mL
Molecular grade water	120 mL

Table 2

Identification and sequence of primers used in the MBLA

Name	Sequence
Mtb 16s Forward	GTGATCTGCCCTGCACTTC
Mtb 16s Reverse	ATCCCACACCGCTAAAGCG
IC MMtmRNA F	CGTCATCCTGGCTAGTTC
IC MMtmRNA R	CTACGGCATTCCCTCAAG
Mtb 16s probe	FAM-AGGACCACGGGATGCATGTCTTGT-BHQ1
IC MMtmRNA pobe	HEX-AGT CCG CTATGT CTC TGC TCG-BHQ1

7. β-mercaptoethanol.
8. Lysing buffer, e.g., FastPrep RNAPRO solution or full FASTRNA Pro Blue Kit with homogenization beads; MP Biomedicals.
9. Chloroform.
10. DNA ase, e.g., TURBO DNA-free, AM190 7 M, Invitrogen.
11. Positive control: *M. bovis* BCG (BCG High & Low concentration) (*see* **Note 2**).
12. Internal control (MM-IC) (*see* **Notes 3** and **4**).
13. Negative control: Water (Molecular grade RNA-ase free water) (*see* **Note 5**).
14. Primers and TaqMan probes, 5′-3′ sequences are found in Table 2:
15. Mastermix suitable for multiplex RT-qPCR, e.g., QuantiTect Multiplex RT-PCR NR Kit, 204845, Qiagen; described in detail in Table 3.
16. RT+, reaction containing reverse transcriptase; RT–, reaction without reverse transcriptase; V, volume; F + R, forward and reverse primer mixture.

Table 3
Components of the reaction master-mix

	RT+ reaction V per reaction (μL)	X number of reactions + 8	RT− reaction V per reaction (μL)	X number of reactions + 8
Quantitect Mastermix	10		10	
Mtb16S F+ R primer mix	0.4		0.4	
Mtb16S-FAM probe	0.2		0.2	
Mm IC tmRNA F + R primer mix	0.4		0.4	
Mm IC tmRNA probe	0.2		0.2	
RT enzyme	0.2		–	
RNA-ase free water	4.6		4.8	
Total volume	16		16	

17. Prepare artificial sputum with 400 mL distilled water, 2.5 g pig stomach mucin (Sigma), 3.0 g diethylene triamine pentacetic acid (DTPA), 2.5 g sodium chloride, 1.1 g potassium chloride, 125 mg each of the 20 essential amino acids. pH is adjusted to 7.0 using 2 M Tris base, water added to 495 mL and the mixture autoclaved. After autoclaving 5 mL egg yolk emulsion is added.

3 Methods

3.1 Sample Preservation

1. Both patient sputum specimens and BCG positive controls in artificial sputum are preserved in GTC and frozen at −80 °C for long term storage or batching purposes.
2. Add 1 mL of homogenized sputum to 4 mL (1:4 ratio) of GTC and freeze at −80 °C until use.
3. If immediate sample processing is required, mix sputum with GTC at 1:4 ratio, leave for 15 min at room temperature. Proceed to RNA extraction.

3.2 Preparation of RT-qPCT Mastermix

1. The mix (QuantiTect Multiplex RT-PCR NR Kit (QT) is supplied in tubes for 200 or 1000 25-μL reactions and includes the reverse transcriptase (RT) enzyme. Aliquot the mix by 200 μL into 1.5 mL RNA-ase free microcentrifuge tube tubes and store at −20 °C. The RT enzyme can be aliquoted at 100 μL and stored at −20 °C until use (*see* **Note 6**).

3.3 Preparation of Primers and Probes

1. Primers and probes are supplied lyophilized and are stable at room temperature (25 °C).
2. Make stock solutions of primers and probes according to the manufacturer's instruction as follows:
   (a) Dissolve primers in RNA-ase-free molecular grade water to prepare a 100 μM stock solution.
   (b) Dissolve probes in the supplied buffer to prepare a 100 μM stock solution (*see* **Note 7**).
3. Prepare primer and probe working solutions:
   (a) Prepare 10 μM of forward + reverse primers solution. Aliquot 160 μL RNA-ase-free water and add 20 μL Forward primer and 20 μL reverse primer from stock.
   (b) Prepare 20 μM probe solution, label RNA-ase-free 1.5 mL microcentrifuge tube tubes with probes names, concentration and the date. Aliquot 80 μL RNA-ase-free water and add 20 μL stock probe.
   (c) Store all working solutions at −20 °C until use and protect probes from direct light during use.

3.4 RNA Extraction (See Note 8)

1. Work within a microbiological safety cabinet and clean the working surfaces with a preparation to remove RNA.
2. Thaw on ice all samples, BCG standards, and the internal control (IC).
3. Prepare corresponding number of homogenization (containing glass micro beads) tubes and label them according to samples' IDs and controls' numbers both on the tube and on the lid. If you number samples then make record with full reference of the patient ID and corresponding tube numbers.
4. Make sure all tubes with samples and controls are properly labeled (name, control concentration or dilution, and date).
5. Add 100 μL internal control suspension into each sample and standard tube. Pipette directly into the liquid and not on the side of the plastic tube. Mix immediately by inverting the tubes. Change pipette tips between the samples.
6. Within the safety cabinet, transfer the tubes into the centrifuge buckets and close the lids.
7. Transfer the buckets to centrifuge and spin at 3000 × *g* for 30 min. Go to **step 10** if you are not working with GTC-preserved samples.
8. After spinning, very gently tip off the GTC supernatant into the corresponding 15 mL tube to be saved at −80 °C or discarded into chemical waste. Do not disturb the cell pellet. If the pellet is detached, spin the tubes again.

9. Pipette 950 μL of the lysing buffer into each sample tube containing the pellet (aim to use the same pipette tip unless you touch the tube).
10. Suspend the pellet with a P1000 tip in the RNApro solution and transfer the suspension to corresponding homogenization tube.
11. Transfer the tubes into a homogeniser and ensure that the tubes are fully pushed in, that the plastic tube holder is engaged with the metal pin and the spokes are located above each tube. On the FastPrep, screw the cap on the top and then close the lid (*see* **Note 9**).
12. Set the FastPrep to programme 6.0 for 40 s. For the Precellys 24, use 40 s at 5000 × *g*.
13. After homogenization, centrifuge the tubes for 5 min at 12,000 × *g* at room temperature.
14. Leave the tube to stand for 5 min following centrifugation.
15. Label fresh 1.5 mL screw cap tubes (RNA-ase free) with the corresponding numbers.
16. Add 300 μL chloroform into each clean tube.
17. From the homogenization tubes, carefully transfer the whole liquid part to the tube containing chloroform using a fine tip pipette (be careful not to transfer bits of sample debris or lysing matrix).
18. Vortex the tubes for 10 s each.
19. Incubate at room temperature for 5 min.
20. Centrifuge at 12,000 × *g* at room temperature for 5 min.
21. Label new 1.5 mL microcentrifuge tube tubes with the corresponding numbers.
22. Carefully transfer the upper aqueous phase to the fresh tubes using 200 μL filter tips, being careful not to transfer any of the interphase or lower layer.
23. Add 500 μL of 100% ice-cold ethanol to each tube.
24. Mix the contents by inverting the tubes five times.
25. Transfer the tubes containing samples to the −20 °C freezer overnight (or to the −80 °C freezer for 15 min if you are doing the whole extraction on the same day) (*see* **Note 10**).
26. Chill the microfuge to at least 12 °C prior to centrifugation (set it at 4 °C) (*see* **Note 11**).
27. Centrifuge the samples at 13,000 × *g* for 20 min.
28. Discard the supernatant using a fine tip pipette tip.
29. Add 500 μL 70% ice-cold ethanol to each tube.
30. Centrifuge as above for a further 10 min.

31. Discard the supernatant using a Pasteur pipette with fine tip. Use a new pipette for every sample.
32. Dry the RNA at 50 °C in a heat block (approximately 20–30 min).
33. Dissolve the extracted RNA in 100 μL RNA-ase-free water at room temperature for 5 min and resuspend the RNA by vortexing for 5 s (*see* **Note 12**).
34. Store the RNA at −80 °C or proceed directly to the DNA ase treatment stage (*see* **Note 13**).

3.5 DNase Treatment (Turbo DNA-Free Kit, Ambion AM1907)

1. When using Turbo DNA-free kit (Ambion AM1907), prepare a master-mix containing Turbo DNase I 10× buffer and DNase I enzyme for the number of samples (plus two extra to avoid pipetting errors). Per sample, add 10 μL buffer and 1 μL DNase enzyme to each sample. For example, a 10 sample master mix would contain 120 μL 10× buffer and 12 μL of DNase enzymes (*see* **Notes 13** and **14**).
2. Add 11 μL of DNA-ase-buffer master-mix to each sample (*see* **Note 15**). Mix by vortexing and then spin briefly (5–10 s at 13,000 *g*) to bring everything to the bottom of the tube. This step eliminates droplets hanging on the tube wall and ensures that all RNA extract is in contact with DNase enzyme.
3. Incubate at 37 °C for 30 min in the hot-block or incubator.
4. After 30 min, add an additional 1 μL of DNase enzyme to each tube. Ensure all enzyme goes in by pipetting up and down x3 and then use the tip to stir around and mix the enzyme with the rest of the mixture.
5. Incubate at 37 °C for a further 30 min in the hot-block or incubator.
6. Thaw the DNase inactivation reagent (white milky substance) 10 min prior to the finish of DNase incubation and keep in the fridge. Resuspend by vortexing.
7. Add 10 μL of DNase inactivation reagent into each RNA extract.
8. Vortex three times during the 5-min incubation step at room temperature.
9. After DNA-ase treatment centrifuge at 13,000 × *g* for 2 min.
10. Label 1.5 mL screw cap tubes with sample ID and extraction date.
11. Set your pipette at 110 μL and carefully pipette off the RNA to a fresh 1.5 mL RNA-ase free tube without touching any of the inactivation matrix.
12. Store the RNA at 4 °C until use if performing the RT-qPCR on the same day.
13. For long term use, store the RNA at −80 °C.

3.6 Reverse Transcriptase—Quantitative PCR (RT-qPCR)

3.6.1 Samples

1. Prepare corresponding number of RNA-ase free 1.5 mL microcentrifuge tube tubes and label them with sample IDs and dates.
2. Dilute all RNA extracts to be used in a 1 in 10 ratio, i.e., add 10 μL of RNA extract into 90 μL of RNA-ase free water. Change tips between samples.
3. Mix well by vortexing for 5 s and briefly spin down to remove any droplets or air bubbles.
4. For every sample, the RNA is analyzed in duplicate, and both neat and 1 in 10 dilution.

3.6.2 Preparing Standard Samples for Standard Curve

1. Take the BCG and IC high concentration RNA extracts and thaw on ice.
2. Arrange 7 RNA-ase free tubes for BCG and 7 tubes for IC standards and label them with the appropriate dilution number. Pipette 90 μL RNAase free water into each tube.
3. Starting with the BCG standards, add 10 μL of the extract to the 90 μL RNA-ase free water in the first tube. Mix well be vortexing for 5 s. This step creates a 1 in 10 dilution.
4. Transfer 10 μL from the first dilution tube into 90 μL RNA-ase free water in the second tube. Mix well by vortexing. Repeat this step until you have diluted the RNA extract up to tube 7. Change tips between each dilution each time prior to transferring sample to the next tube.
5. Repeat **steps 3** and **4** for IC standards.

3.6.3 Master Mix Preparation

Mastermix is a solution of PCR reagents sufficient for all samples and standards to be amplified. Each RNA sample and its decimal dilution, and standard will be amplified in duplicate. Calculate the master-mix volume taking into account the total number of reactions and including eight extra reactions to maintain the minimum liquid level when using a pipetting robot to set up the reactions. The composition of master-mix is outlined in Table 2.

1. Use a pipetting robot, e.g., the Qiagility to pipette master mix and RNA into PCR reaction tubes (*see* **Note 16**).
2. After each use and/or prior to pipetting, decontaminate the pipetting robot and appropriate racks and holders by UV light for 15 min.
3. Defrost the PCR reagents in the clean room.
4. UV decontaminate an aliquot of molecular grade water.
5. Instruct the Robot to add 16 μL master mix and 4 μL sample RNA into the reaction tubes (0.2 mL thin-walled PCR tubes for 36-well rotor or 4-strip cap tubes for 72-well rotor).

6. Each sample and control should be run in duplicate in RT-qPCR (RT+, containing the RT enzyme) and in a single reaction for master mix without reverse transcriptase (RT−) (*see* **Notes 17** and **18**).
7. In absence of the robot, use the 72-well pipetting plate to manually pipette the master mix and RNA (*see* **Note 19**).
8. Carefully close the lids and transfer the tubes into the corresponding rotor of the RotorGene Q.

3.6.4 Thermocycler Set Up Using RotorGene Q

1. When the reaction tubes are in the appropriate rotor, lock them in the position with the locking ring. Place the rotor into RotorGene, click into position and close the lid of the instrument (*see* **Note 20**).
2. Switch on the RotorGene Q instrument.
3. Switch on the computer and open the RotorGene Q software.
4. Name the operator and write notes defining your assay.
5. Define the reaction volume, 20 μL.
6. Define reaction conditions by clicking on the edit profile button:
   (a) Hold at 50 °C, 30 min [this is the reverse transcription].
   (b) Hold at 95 °C, 15 min [this activates the Taq polymerase].
   (c) Cycling, 40 cycles of: 94 °C, 45 s not acquiring, 60 °C, 60 s acquiring at Green and Yellow (click on the acquiring button to edit fluorescence channels you need for the reaction).
   (d) Click OK, which takes you back to the summary window.
   (e) Check the Gain optimization button at first acquisition and leave settings as default.
   (f) Click Next which takes you to the summary of reaction conditions.
   (g) If reaction conditions are correct press the "Start Run" button. You will be asked to save the reaction and after saving the reaction will start and open the sample sheet for sample information input.
   (h) Complete the sample sheet and click the finish button at the bottom of the sheet.

3.6.5 Result Interpretation and qPCR Output Data Analysis

Careful PCR data interpretation is crucial for correct Mtb bacterial load assessment and treatment monitoring. Amplification data for Mtb samples and BCG and IC controls must be evaluated at all times. Table 4 illustrates how MBLA results are interpreted relative to the internal control:

Table 4
Interpretation of results

Target (Mtb/BCG)	IC	Result
+	+	+
+	–	+[a]
–	+	–
–	–	Invalid

+, positive amplification in RT-qPCR
–, negative amplification in RT-qPCR
Invalid, both target and IC are negative, no amplification detected
[a]Positive for Mtb but the result cannot be used for quantitative analysis or data normalization as IC was not detected

3.6.6 Troubleshooting

1. If no amplification is detected for the internal control (Yellow channel), ascertain that the internal control was added into the samples by revisiting your extraction checklist to confirm that the internal control addition step was performed.
2. Ascertain that your qPCR master mix was correctly composed to include both target (Mtb or BCG) and IC primers and probes.
3. If the answer to 1 and 2 is YES, IC should be positive at least in some of the samples and BCG controls extracted in the same batch.
4. If the answer to 1 is NO, IC will be negative in all samples and BCG controls. Repeat the RNA extraction using sampled spiked with IC.
5. If the answer to 2 is NO, then remake the master mix and repeat the RT-qPCR.

3.7 Construction of Standard Curves for the MBL Assay

3.7.1 Principle

The Molecular Bacterial Load assay (MBLA) monitors the molecular load of Mtb cells in sputum samples and can provide accurate information on bacterial response (decline) to antimicrobial treatment. The concentration of TB is calculated from 16S rRNA present in the sample. The principle of the MBL assay is absolute quantification based on a standard curve consisting of a set of RNA templates with known concentration. The standard curve is used to calculate the Mtb concentration of an unknown clinical sample.

A separate standard curve is used for Mtb quantification and a separate curve is used for the quantification of mycobacterial internal control (IC). Standard curves must be constructed for each real-time PCR instrument, and can be then incorporated into PCR runs and adjusted according to the high and low BCG controls in the run.

3.7.2 Standard Curves Construction

1. Use RNA extracted from standard material as detailed (*see* **Notes 2** and **3**), i.e., *M. bovis* BCG culture and MM-IC cell suspensions with estimated concentrations of 10^8 CFU/mL or greater. Work on ice.
2. Prepare 7 1.5 mL RNase-free tubes and label them with appropriate name and dilution number (e.g., 1–7). Dilute the extracted RNA decimally to create a series of standards. Add 10 μL of extracted RNA into 90 μL of RNase-free water, mix by vortexing for 5 s. Prepare 7 decimal dilutions. Change tips between each dilution.
3. Set up the RT-qPCR mastermixes as outlined above in chapter "RT-qPCR" using appropriate primers and probes combinations.
4. The standards are amplified in duplicates (along with the samples or on their own). Use 4 μL of RNA dilution and 16 μL of mastermix per reaction.
5. For each type of RotorGene Q rotor, i.e., 36-well or 72-well, create a separate standard curve. Note: the reaction volume, type of reaction tubes and rotor can influence the PCR performance and fluorescence reading, and hence, cause minor variations in fluorescence signal and the resulting C_T values.
6. In RotorGene Q software, label the standards in sample sheet and assign them corresponding concentration and units, 10E9 copies/mL for the neat BCG RNA, 10E8 for the first 1 in 10 dilution, etc.

3.7.3 Standard Curve Data Analysis

A standard curve can be prepared in a separate run for the use with RotorGene Q and it can be further incorporated in data analysis of samples with unknown bacterial load.

1. Analyze the amplification curves in appropriate fluorescence channel, i.e., green channel for Mtb (FAM labeled probe), yellow channel for IC (VIC or HEX labeled probe).
2. Set the fluorescence threshold to 0.01 and examine the curves in exponential view and then in logarithmic mode. Use the same threshold for both Mtb and IC channels. Note: the threshold can be set to 0.02 if this improves the efficiency of the reaction.
3. Go to "Analysis" option and select the channel and sample sheet you are going to analyze.
4. Click on "Slope correct" in order to minimize the fluorescence fluctuations.
5. When standards and their respective concentrations are assigned in the sample sheet, the analysis software will automatically populate a standard curve.

6. Examine the parameters of the standard curve. The parameters are:
   (a) Slope (M), informs on assay efficiency.
   (b) Correlation coefficient (R^2), assesses the standard curve linearity and the dynamic range (or limits of quantification).
   (c) Intercept, shift in C_T value on the y axis.
7. The PCR efficiency can be evaluated by the parameters of standard curve. The equation for an ideal standard curve and a 100% amplification efficiency ($E = 1$) is:

   $$C_T = \text{slope} \times \text{Log(concentration)—intercept.}$$

   or

   $$C_T = -3.32 \times \text{Log (concentration)—intercept.}$$

   Aim for the efficiency of 90–100%, i.e., $E = 0.9$–1.0. The efficiency can be calculated from the slope of the standard curve using the equation:

   $$E = 10^{-1/-3.32}-1$$

8. Very high or too low RNA concentrations in the RT-PCR reaction can cause fluctuations in reverse transcription and subsequent PCR efficiency. These result in outlying C_T values. Outlier C_T values can be also caused by the errors in pipetting, dilutions' preparation, insufficient homogeneity of a PCR mastermix, evaporation during reaction, improperly placed rotor.
9. Carefully consider removal of the outliers.

3.7.4 Importing the Standard Curve for MBLA Analysis

MBLA is an assay for absolute quantification of an unknown target. Every RNA extraction includes the extraction from high and low BCG positive controls along with the preadded internal control. The BCG high and low controls contain a known concentration of BCG. Thus the quantification can be adjusted to these samples.

1. In RotorGene Q software, label the BCG high and low controls as standards and assign them appropriate concentration (10^8 or 10^5 etc.) and units (copies/mL).
2. When amplification of a series of unknown samples to be analyzed is finished, set the fluorescence threshold to 0.02 (i.e., the same value as the threshold for the standard curves).
3. Select "Slope correct" option.
4. Analyze one channel and one sample sheet at a time, i.e., the amplification in green channel for Mtb samples.
5. In the Analysis mode, select log view of the amplification curves.
6. Click on the amplification curves window. An option for standard curve import will appear in the right-hand side of the screen.

7. Import the standard curve from a previous run. Note: make sure you can localize the standard curve run file and that you recognize the quantification parameters, i.e., slope (M), intercept, and R^2.
8. The software now offers you the option of adjusting the reaction efficiency to the current run. The efficiency will be adjusted based on the high and low BCG controls, which you had assigned as standards. Select adjust and continue.
9. The software now automatically populates the standard curve into your current run and calculates the concentrations of the unknown samples. The samples will appear plotted on the standard curve according to their C_T values and calculated concentrations. The calculated concentrations and standard deviations for the duplicate samples will appear in the results table. The calculated concentrations are in estimated CFU/mL units (eCFU/mL).
10. Examine the C_T values and calculated concentrations; focus on the standard deviation between duplicate samples. Compare the C_T values and calculated concentrations of neat and decimally diluted RNA.
11. If the difference in C_T values is around 3.32 or the calculated concentration difference is tenfold between the neat and 1 in 10 diluted samples, there was no inhibition of the reaction (usually coming from residual DNase stop solution) and both concentrations of neat and diluted RNA indicate the Mtb concentration in the sample.
12. When calculating the Mtb concentration in the sample, always consider the dilution used for PCR or for the RNA extraction, i.e., multiply the result by ten for a tenfold diluted sample.
13. Record the results in appropriate results sheet.

4 Notes

1. Preparation of GTC working solution:

   Work in the fume cabinet wearing lab coat, gloves, and safety goggles. Carefully weigh and transfer 200 g GTC powder into the 500 mL plastic container. Measure 120 mL molecular grade water and transfer to 500 mL plastic container containing GTC. Shake thoroughly and then incubate overnight at 37 °C to dissolve the GTC. You may need to shake again and allow the remaining GTC powder another 30 min to dissolve the next day. Add 40 mL 1 M Tris–HCl pH 7.5 and adjust the volume to 396 mL with molecular grade water. When the GTC solution is completely clear, add 4 mL β-mercaptoethanol and mix well by shaking. Aliquot 4 mL or 8 mL into 15 mL Falcon tubes and freeze immediately at −80 °C (Do not leave more than 2 h on the bench).

Adding sample to GTC and freezing at −80 °C preserves RNA in the sample for a long period of time. This allows sample batching and simplifies sample processing for routine clinical laboratories. GTC also inactivates *M. tuberculosis* reducing risk of infection for the operators.

2. Preparation of *M. bovis* BCG High and Low Positive controls.

   The positive controls are made from a log phase culture of *M. bovis* BCG (BCG). The culture is prepared by inoculating BCG into 7H9 medium (with 10% OADC or ADC growth supplement) and incubating it at 37 °C for up to 3 weeks. The bacterial load of positive controls is confirmed by plating on 7H10 agar for CFU counts and qPCR confirmation. Make **BCG high (H) positive control** by diluting the culture into artificial sputum to a final concentration of 10^7–10^8 CFU/mL and preserve in GTC at a ratio of 1:4, e.g., to a total volume of 5 mL in 15 mL Falcon tubes. Store the controls at −80 °C until use. One or two BCG H controls should be co-extracted in every patient sample RNA extraction batch. The **BCG low (L) positive control** is made by diluting the BCG culture into artificial sputum to a concentration of 10^3 to 10^4 CFU/mL and then preserved in GTC at a ratio of 1:4 to a total volume of 5 mL in 15 mL Falcon tubes. Store the controls at –80 °C until use. One or two BCG L controls should be co-extracted in every patient sample extraction batch.

3. Preparation of the Internal Control (IC).

   The Internal (extraction) control is prepared from a log phase of *Mycobacterium marinum* cultured in 7H9 medium (without growth supplement ADC or OADC) at 30 °C. Centrifuge a fraction of the culture at 3000 × *g* for 30 min (or 10 min 10,000 × *g* for a benchtop centrifuge), discard the supernatant and freeze the cell pellets at –80 °C. Use a different fraction of the same *M. marinum* culture to determine its concentration (CFU/mL) by inoculating 7H10 agar and incubating at 30 °C for 3–5 days. Thaw the frozen pellets at room temperature and resuspend in GTC supplemented with 1% β-mercaptoethanol solution to 10^5 CFU/mL. Store at –80 °C for long-term use. At the time of sample RNA extraction, spike each sample and BCG positive control with 100 μL of the internal control suspension. The final concentration of internal control in the sample should be 10^4 CFU/mL.

4. It is essential that the internal control is carefully managed as there is a risk of cross-reaction if the concentration of the internal control is greater than 10^4 CFU/mL.

5. Molecular grade water. Keep the water sterile at room temperature. To prevent any contamination, it is recommended to aliquot small amounts into RNase free vessels to work with. *The water aliquot used for RNA extraction and RT-PCR setup should also be used as a no template control (NTC) or negative control for PCR.*
6. Before setting up a PCR, work out how much QT PCR mastermix needed and take out the exact number of aliquots to use. Avoid multiple freeze–thaw cycles of the QT mix.
7. Store the oligonucleotides at −20 °C and protect the probes from direct light by using colored nontransparent tubes or wrapping the tubes in aluminum foil during use.
8. All procedures must be completed in a microbiological safety cabinet until the organisms are confirmed to be killed. The lysis kills Mtb, and the rest of the extraction can continue in a biosafety II lab. Clean the outsides of all tubes with disinfectant before they are transferred into the biosafety II lab.
9. FastPrep instrument or alternative bead beater will increase the yield.
10. Sometimes the tube will freeze at −80 °C if there is a lot of salt in the sample. Allow to thaw at room temperature, for approximately 5 min.
11. Orientate the tubes in the centrifuge so that hinged back of the tube sits on the outside diameter of the rotor. This way you will know where the pellet is sitting when you remove the supernatant.
12. If the reconstituted RNA is viscous and does not resuspend, add a further 100 μL of water.
13. The tube top colors change, so make sure you read the labels of the tubes to ascertain that you are adding the right reagent.
14. It might be necessary to add 15 μL in tubes where more DNA is anticipated, e.g., in baseline samples and BCG high standard.
15. The master mix accounts for 16 μL of the reaction volume. Adding 4 μL of sample (RNA extract) makes the total volume of reaction 20 μL.
16. Manual pipetting can be used if only few samples, e.g., ≤10 samples, are being processed.
17. Once qPCR without any added RT enzyme (RT−) has been done on samples and the result is negative, e.g., there is no DNA present, there is no need to repeat this for subsequent runs of the same samples.
18. The following number of reactions per run can be calculated: 72-well rotor: Samples 20 × 2 RT+, 20 × 1 RT−, 2 × 2 BCG H

& 2 × 2 BCG L RT+; 1 × 1 BCG H & 1 × 1 BCG L RT, 1 × 2 Water (NTC) RT+, 1 × 1 Water RT− = 71 reactions in total.

36-well rotor: Samples 9 × 2 RT+, 9 × 1 RT−, 1 × 2 BCG H & 1 × 2 BCG L RT+; 1 × 1 BCG H & 1 × 1 BCG L RT−, 1 × 2 Water (NTC) RT+, 1 × 1 Water RT− = 36 reactions in total.

19. All samples and BCG controls are run, neat and its 1 in 10 dilution. Standard curve samples do not have to be run every time. The standard curves can be imported from a previous run.
20. Make sure the names in the sample follow the order of the reaction tubes on the rotor.

References

1. Honeyborne I, McHugh TD, Phillips PPJ et al (2011) Molecular bacterial load assay, a culture-free biomarker for rapid and accurate quantification of sputum mycobacterium tuberculosis bacillary load during treatment. J Clin Microbiol 49:3905–3911
2. Honeyborne I, Mtafya B, Phillips PPJ et al (2014) The molecular bacterial load assay replaces solid culture for measuring early bactericidal response to anti-tuberculosis treatment. J Clin Microbiol 52(8): 3064–3067
3. Bowness R, Boeree MJ, Aarnoutse R et al (2014) The relationship between Mycobacterium tuberculosis MGIT time to positivity and cfu in sputum samples demonstrates changing bacterial phenotypes potentially reflecting the impact of chemotherapy on critical sub-populations. J Antimicrob Chemother 70(2):448–455. doi:10.1093/jac/dku415
4. Mukamolova GV, Turapov O, Malkin J et al (2010) Resuscitation-promoting factors reveal an occult population of tubercle bacilli in sputum. Am J Respir Crit Care Med 181:174–180

Chapter 6

Defining Diagnostic Biomarkers Using Shotgun Proteomics and MALDI-TOF Mass Spectrometry

Jean Armengaud

Abstract

Whole-cell MALDI-TOF has become a robust and widely used tool to quickly identify any pathogen. In addition to being routinely used in hospitals, it is also useful for low cost dereplication in large scale screening procedures of new environmental isolates for environmental biotechnology or taxonomical applications. Here, I describe how specific biomarkers can be defined using shotgun proteomics and whole-cell MALDI-TOF mass spectrometry. Based on MALDI-TOF spectra recorded on a given set of pathogens with internal calibrants, *m/z* values of interest are extracted. The proteins which contribute to these peaks are deduced from label-free shotgun proteomics measurements carried out on the same sample. Quantitative information based on the spectral count approach allows ranking the most probable candidates. Proteogenomic approaches help to define whether these proteins give the same *m/z* values along the whole taxon under consideration or result in heterogeneous lists. These specific biomarkers nicely complement conventional profiling approaches and may help to better define groups of organisms, for example at the subspecies level.

Key words Bacterial proteome, Biomarkers, Dereplication strategy, Diagnostic, High-throughput proteomics, Pathogens, Proteogenomics, Screening, Spectral count, Tandem mass spectrometry, Whole-cell MALDI-TOF

1 Introduction

Mass spectrometry is today a widespread tool for bacterial identification because of its simplicity of use, its reproducibility and sensitivity, and its startling quick-time response [1–4]. Different approaches have been developed for specific and highly sensitive searches for the presence of a given pathogen in a sample. Some are eminently complex but really gripping, as exemplified by the detection of *Bacillus anthracis* spores by a combination of immunocapture, trypsin and Glu-C proteolysis, liquid chromatography, and tandem mass spectrometry [5]. In this case, proteotypic peptides of a given protein biomarker, the small acid-soluble spore protein-B, are monitored through specific fragment ions produced by fragmentation of the isolated peptide in a collision cell with the

Kimberly A. Bishop-Lilly (ed.), *Diagnostic Bacteriology: Methods and Protocols*, Methods in Molecular Biology, vol. 1616, DOI 10.1007/978-1-4939-7037-7_6,

multiple reaction monitoring mass spectrometry mode. Such analysis may also give resistance and virulence information on clinically relevant pathogens [6]. Other approaches could be considered "without any a priori," as the methodology applies to whatever organism is present in the sample (reviewed in [7]). Simple mass spectrometry or tandem mass spectrometry of peptides generated by proteolysis of the proteins present in the sample and comparison to comprehensive genomic/proteomic databases can be a possible approach [8, 9], but the method is relatively time-consuming and deserves further improvements for its generalization. Top-down identification of bacterial proteins by high-performant tandem mass spectrometry and search of homologs is also an alternative but until now it has been difficult to perform [10]. For now, Matrix-Assisted Laser Desorption Ionization-Time Of Flight (MALDI-TOF) mass spectrometry on whole bacterial cells has been shown to be a reliable method [11]. Its principle consists in recording by mass spectrometry the masses of the different abundant proteins from a sample and comparing the resulting profile with a database encompassing all the profiles of known organisms previously recorded in the same experimental condition [12]. Whole-cell MALDI-TOF requires little handling: deposition of a single plate-grown colony mixed with a chemical matrix on the MALDI-TOF target plate, and then automatic acquisition of a MALDI-TOF *m/z* spectrum and database comparison. Characteristic ions in the MALDI TOF mass spectra recorded from bacteria have been associated with low molecular weight, basic and abundant proteins [13, 14]. The most conserved *m/z* peaks among different bacteria may define biomarkers which can then be characterized to better assess the reliability of species discriminants [15].

The method to define biomarkers detectable by MALDI-TOF mass spectrometry outlined here is, in my experience, relatively straightforward and does not require specific expertise, while other elegant methods relying on expert top-down mass spectrometry are time-consuming and involve a tandem mass spectrometer with specific fragmentation mode. This method relies on proteogenomics, the alliance of genomics and proteomics, as some biomarkers may be wrongly annotated in the genome or proteome databases [16, 17]. It is derived from previous methodologies applied for the search of biomarkers directly from the protein database from different bacteria, such as *Escherichia coli* [18] and *Lactobacillus plantarum* [19]. Using the method detailed below, biomarkers have been proposed for screening new environmental strains from the *Ruegeria* genus [20] and the *Roseobacter* clade [21], or pathogens such as *Neisseria meningitidis* [22] or *Francisella tularensis* [23].

2 Materials

1. Bacterial strains to be analyzed (*see* **Notes 1–4**).
2. Appropriate solid media on which to grow bacteria, or alternatively liquid media.
3. Sterile plastic inoculating loops.
4. 1.5 mL centrifuge tubes.
5. Clean scalpel or razor blade.
6. High speed centrifuge.
7. Electrophoresis system.
8. Sonicator.
9. Heating block—37 °C, 56 °C, 95 °C.
10. Orbital shaker.
11. SpeedVac.
12. MALDI-TOF Biflex IV (Brucker Daltonics) MALDI-TOF mass spectrometer, or equivalent material.
13. LTQ Orbitrap XL (ThermoFisher) tandem mass spectrometer coupled to an UltiMate 3000 LC system (Dionex), or equivalent material.
14. Reverse-phase Acclaim PepMap100 C18 µ-precolumn (5 µm, 100 Å, 300 µm i.d. × 5 mm, Dionex-ThermoFisher) and nanoscale Acclaim PepMap100 C18 capillary column (3 µm, 100 Å, 75 µm i.d. × 15 cm, Dionex), or equivalent materials.
15. Polyacrylamide gels: e.g., 4–12% Bis-Tris gradient 10-well gels. Store at 4 °C.
16. Matrix solution: typically, alpha-cyano-4-hydroxycinnaminic acid (HCAA) prepared in 50% (vol/vol) aqueous acetonitrile containing 2.5% (vol/vol) trifluoroacetic acid.
17. 4× Lithium dodecyl sulfate (LDS) buffer (40% glycerol, 4% LDS, 4% Ficoll-400, 800 mM triethanolamine–HCl pH 7.6, 0.025% phenol red, 0.025% Coomassie G250, and 2 mM EDTA disodium).
18. 80% (vol/vol) aqueous trifluoroacetic acid.
19. PAGE running buffer: Dilute 100 mL of 20× MES buffer (1 M 2-(N-morpholino)ethanesulfonic acid, 1 M Tris Base, 2.0% sodium dodecyl sulfate, 20 mM EDTA, pH 7.3) in 1.9 L of water. Store at 4 °C.
20. Coomassie Blue stain.
21. Deionized water.
22. Pure acetonitrile.
23. Destain solution: Mix 10 mL of methanol with 10 mL of 50 mM ammonium bicarbonate.

24. Dehydration solution: Mix 10 mL of acetonitrile with 10 mL of 50 mM ammonium bicarbonate.
25. Reduction solution: Weigh 38.5 mg of dithiothreitol in a 15-mL centrifuge tube. Add 10 mL of 50 mM ammonium bicarbonate.
26. Alkylation solution: Weigh 102 mg of iodoacetamide in a 15-mL centrifuge tube. Add 10 mL of 50 mM ammonium bicarbonate. Store in the dark.
27. Enzyme solution: Reconstitute lyophilized sequencing-grade trypsin to a final concentration of 0.1 μg/μL in 0.01% trifluoroacetic acid (*see* **Note 5**).
28. Digestion solution: 16 μL of 50 mM ammonium bicarbonate, 2 μL of 0.1% ProteaseMax, and 2 μL of 0.1 μg/μL trypsin (20 μL total volume) per gel piece. Keep on ice until use.
29. Trifluoroacetic acid: 5% stock solution in water. Prepare 0.5% and 0.1% solutions in water.
30. Solvents for liquid chromatography: 0.1% formic acid (Solvent A) and 0.1% formic acid, 80% acetonitrile (Solvent B) in water.
31. Microsoft Excel software.

3 Methods

3.1 Colony Transfer and Treatment

1. Transfer with a plastic, sterile inoculation loop a visible amount of bacterial biomass sampled from a large colony grown on agar plate directly to the sample spot of the MALDI-TOF target (*see* **Note 6**).
2. Overlay the deposited material with the appropriate amount of matrix solution.
3. Allow to dry for 10 min prior to inserting the target plate into the MALDI-TOF mass spectrometer.
4. For shotgun proteomics, prepare 1× LDS buffer by diluting a 4× solution with water.
5. Transfer with the inoculation loop a visible amount of bacterial biomass sampled from a large colony grown on agar plate directly into 100 μL of sterile water in a 1.5 mL Eppendorf tube.
6. Centrifuge for 5 min at 6000 × *g* at room temperature.
7. Remove the supernatant, suspend the cells directly in 50 μL of 1× LDS buffer, and sonicate briefly for 1 min with an ultrasonic probe to dissolve the cellular pellet.
8. Heat the sample at 95 °C for 5 min in a heating block.
9. Centrifuge the tube briefly to obtain the soluble sample for SDS-PAGE.

10. Load 20 μL of LDS boiled sample onto a polyacrylamide gel, e.g., 4–12% Bis-Tris gradient 10-well gels, with molecular weight standards on the well beside.
11. Run at 200 V with 1× MES buffer for approximately 20 min until the entire sample has entered into the gel and the proteins resolved enough to separate high and low molecular weight molecules.
12. After migration, rinse the gel with water, stain with coomassie blue for 30 min, and rinse well with water.
13. Excise with a clean scalpel or razor blade the low molecular weight proteins (below 20 kDa) as a single band of roughly 250 μL volume (if possible) or alternatively as several bands from high to low molecular weights.
14. Transfer the polyacrylamide band to eppendorf tube.
15. Destain for 1 min with 200 μL of methanol–ammonium bicarbonate under shaking at 500 rpm, and discard the fluid. Repeat this step once.
16. Dehydrate for 5 min with 200 μL of acetonitrile:ammonium bicarbonate under shaking at 600 rpm, and discard the fluid.
17. Dehydrate for 1 min with 200 μL of pure acetonitrile under shaking at 600 rpm, and discard the fluid.
18. Dry in a SpeedVac for 2–5 min.
19. Rehydrate the gel piece(s) with 100 μL of reduction solution; incubate for 20 min at 56 °C, shaking at 500 rpm. Discard the fluid.
20. Add 100 μL of alkylation solution; incubate for 20 min at room temperature in the dark. Discard the fluid.
21. Wash with 400 μL of deionized water; shake for 1 min at 600 rpm; discard the fluid. Repeat this step once.
22. Dehydrate with 200 μL of acetonitrile:ammonium bicarbonate, shake for 5 min 600 rpm, and discard the fluid.
23. Dehydrate with 200 μL of pure acetonitrile, shake for 1 min at 600 rpm, and discard the fluid.
24. Dry in a SpeedVac for 2–5 min.
25. Rehydrate the gel piece(s) with 20 μL of enzyme solution, and incubate for 20 min on ice. For increasing peptide recovery, use ProteaseMax detergent as recommended (Hartmann et al., 2014). Remove excess liquid and incubate for 4 h at 37 °C.
26. Transfer the solution to a clean tube and add 5 μL of 5% trifluoroacetic acid. If the recovered volume is less than 50 μL, which is often the case for larger gel pieces, add 0.1% trifluoroacetic acid equivalent to the lost volume, shake for 5 min at 500 rpm, and pool the solution with the previously recovered volume. Peptide solutions that will not be analyzed immediately should be frozen and stored at −20 °C (*see* **Note 7**).

3.2 Experimental Data Acquisition by Mass Spectrometry

3.2.1 Acquisition of MALDI-TOF MS Spectra

1. Calibrate the MALDI-TOF mass spectrometer using standard settings in the range of 2000–20,000 *m/z* with an appropriate mixture of protein calibrants deposited in an external calibration spot located near the sample deposits to be measured (*see* **Note 8**). In order to improve the accuracy of the *m/z* peaks of interest, internal calibration is recommended (*see* **Note 9**).
2. Acquire four independent MALDI-TOF spectra on four different biological replicates, each summing at least 150 consecutive laser shots. Repeat this measurement for each condition or for each strain to be assayed.

3.2.2 Low Molecular Weight Shotgun Proteomics

1. Load 10 μL of the resulting peptide mixture onto a nanoLC-MS/MS high resolution system, e.g., LTQ Orbitrap XL (Thermo) coupled to an UltiMate 3000 LC system (Thermo) equipped with a reverse-phase Acclaim PepMap100 C18 μ-precolumn (5 μm, 100 Å, 300 μm i.d. × 5 mm, Thermo) followed by a nanoscale Acclaim PepMap100 C18 capillary column (3 μm, 100 Å, 75 μm i.d. × 15 cm, Thermo).
2. Resolve the peptides over a 60 min linear gradient from 5 to 60% solvent B using a flow rate of 0.3 μL/min. Record full-scan mass spectra over the 300–1800 *m/z* range and MS/MS on the most abundant precursor ions with 60 s dynamic exclusion of previously selected ions. Repeat the measurements for three independent biological replicates.

3.3 Mass Spectrometry Data Processing

3.3.1 MALDI-TOF Data Processing

1. Identify the internal calibration markers in each of the four MALDI-TOF spectra. Based on these peaks, recalibrate the whole spectra.
2. Once recalibrated, export the most intense *m/z* peaks (e.g., the 100 most intense *m/z* peaks) in an excel sheet.
3. Identify the most intense *m/z* peaks which are common in the four spectra, taking an error tolerance of ±200 ppm if internal calibration is done or ±400 ppm when relying on external calibration only.
4. Determine the mean *m/z* value for each of the conserved peaks. Fig. 1 shows an example of four MALDI-TOF spectra acquired on the *Ruegeria lacuscaerulensis* ITI-1157 bacterium. The 4 *m/z* values for a potential biomarker exhibiting an intense signal are indicated, giving an average *m/z* value at 9989.6 (internal calibration) or 9990.2 (external calibration).

3.3.2 Processing of Low Molecular Weight Shotgun Data

1. Search MS/MS spectra against a protein database comprising all the proteins annotated in the genome reference using the following parameters: 2 (maximum number of missed cleavages), 5 ppm (mass tolerance for the parent ion), 0.5 Da (mass tolerance for the product ions), carbamidomethylated cysteine

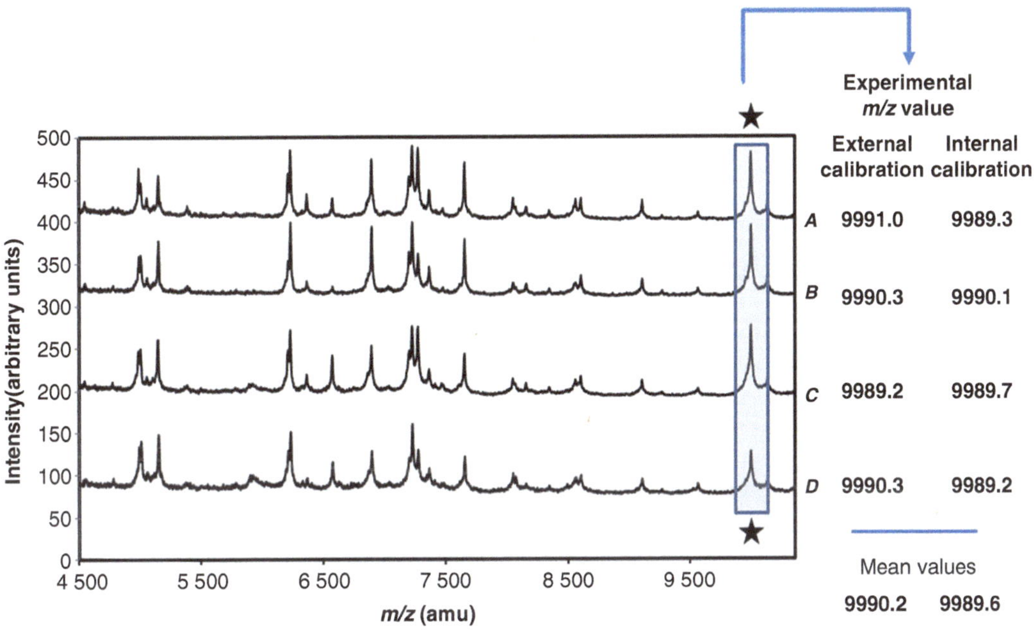

Fig. 1 MALDI-TOF mass spectra of a representative strain of *Ruegeria*. MALDI-TOF positive ion mass spectra have been recorded with *R. lacuscaerulensis* ITI-1157 bacteria with a Biflex IV MALDI-TOF instrument (Bruker Daltonics). Four representative spectra have been overlaid with a baseline shift to help direct visual comparison. The signal is shown in the 4500–10,500 *m/z* range. The intensity is expressed as arbitrary units. A conserved signal at *m/z* 9990 has been *highlighted*. The *m/z* values obtained with external calibration or with internal calibration are indicated for the four spectra labeled A, B, C, and D from *top* to *bottom*, as well as the mean *m/z* values

residues (fixed modification), oxidized methionine residues (variable modification).

2. Validate the proteins with at least two identified peptides and take into account polypeptides detected with only a single peptide as soon as their molecular weight is below 12 kDa.

3. Extract their spectral counts (total number of MS/MS spectra identified per protein), normalize their abundance by dividing their spectral counts per their molecular weights (NSAF, normalized spectral abundance factor), calculate their isoelectric point, and rank the identified proteins along their predicted molecular weights. As an example, Table 1 shows the proteins identified by shotgun proteomics data acquired on the *R. lacuscaerulensis* ITI-1157 bacterium in the range 9–11 kDa [20]. The most abundant protein, as judged from the number of MS/MS spectra recorded which is 159 and the NSAF percentage estimated at 3.7%, is the DNA-binding protein HU, predicted from the genome annotation as having a theoretical molecular weight of 10019.6 Da. This basic and abundant protein should logically explain the *m/z* peak observed in the MALDI-TOF spectra as described here below.

Table 1

List of polypeptides detected for *Ruegeria lacuscaerulensis* ITI-1157 by shotgun proteomics in the range 9–11 kDa

Protein NCBI reference	Functional annotation	Theoretical isoelectric point	Theoretical molecular weight (Da)[a]	Number of spectra	%NSAF
ZP_05786428.1	Ribosomal protein L24	10.2	10924.6	28	0.61
ZP_05786824.1	Chaperonin GroS	4.8	10881.4	62	1.35
ZP_05785138.1	Integration host factor (beta)	9.6	10818.3	24	0.52
ZP_05786402.1	Ribosomal protein L23	9.7	10706.4	6	0.13
ZP_05787274.1	Preprotein translocase (YajC)	9.3	10569.5	33	0.74
ZP_05786405.1	Ribosomal protein S19	10.0	10479.0	46	1.04
ZP_05787717.1	Conserved hypothetical protein	6.1	10377.6	4	0.09
ZP_05786156.1	Ribosomal protein S15	10.1	10294.9	10	0.23
ZP_05786048.1	Conserved hypothetical protein	4.5	10271.6	11	0.25
ZP_05786969.1	Conserved hypothetical protein	4.9	10271.6	36	0.83
ZP_05785873.1	**DNA-binding protein HU**	**9.4**	**10119.6**	**159**	**3.71**
ZP_05787093.1	Hypothetical protein SL1157_2263	8.0	9810.1	8	0.19
ZP_05785902.1	YCII-related protein	4.5	9502.8	3	0.07
ZP_05784961.1	Conserved domain protein	5.3	9395.5	4	0.10
ZP_05784629.1	Putative lipoprotein	6.5	9258.6	27	0.69

The most abundant protein is indicated in bold letters

[a]The theoretical molecular weights have been calculated taking into account the polypeptide sequences from the NCBI protein database, which correspond to the direct translation of the corresponding genes

3.3.3 Identifying Proteins Responsible of MALDI-TOF m/z Signals with the Help of the Low Molecular Weight Shotgun Data

1. In the excel list of identified low-molecular weight proteins, add the genome annotated protein sequence in regard to each protein identifier.
2. Insert four columns in the excel list and calculate the four possible average molecular weights (*see* **Note 10**) taking into account whether the methionine is removed and whether acetylation occurs at the resulting N-terminus (*see* **Note 11**). As exemplified in Fig. 2, the four molecular weights have been calculated for the DNA-binding protein HU (ZP_05785873.1) from the

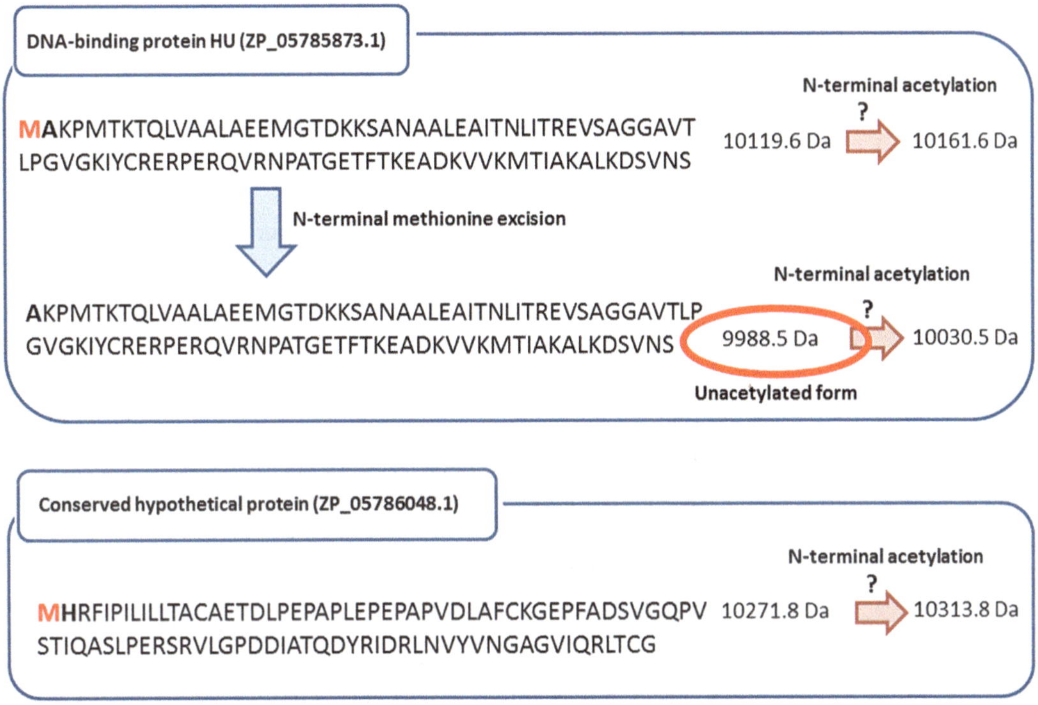

Fig. 2 Molecular weight calculation for two proteins detected by shotgun proteomics as possible biomarkers. The sequences of the DNA-binding protein HU and a conserved hypothetical protein detected by shotgun proteomics from *R. lacuscaerulensis* ITI-1157 bacteria total are shown as indicated in the National Center for Biotechnology Information (NCBI) protein database. The possible posttranslational maturations (N-terminal methionine removal, N-terminal acetylation) are indicated as well as the resulting average molecular weights. The 94-amino acid matured HU protein exhibits an average molecular weight of 9988.5 Da. Its monocharged positive ion, expected at *m/z* 9989.5 explains the signal observed in Fig. 1

R. lacuscaerulensis ITI-1157 bacterium. The annotated protein has a length of 95 amino acids and an average molecular weight of 10119.6 Da. It is reasonable to predict that the initial methionine is removed in the cellular context as the second residue of the polypeptide is an alanine. Therefore, the expected molecular weight of the resulting 94-amino acid matured protein is 9988.5 Da. If these two polypeptides are acetylated, then the molecular weights could be 10161.6 Da or 10030.5 Da, respectively. Another protein from the same organism, ZP_05786048.1 which has been annotated as "conserved hypothetical protein," has a predicted average molecular weight of 10271.8 Da. If acetylated, the resulting molecular weight is 10313.8 Da. In this case, because of the presence of a bulky residue, namely histidine, at the second position of the polypeptide, its initial methionine should not be removed. Therefore, the two molecular weights calculated for this improbable maturation, whether or not acetylated, are not further taken into consideration.

3. Transform these molecular weights into the two possible *m/z* values taking into account the possible occurrence of a monocharged ion and a double charged ion (*see* **Note 12**).
4. Compare with excel the list of theoretical *m/z* values with the list of experimental *m/z* values detected on the MALDI-TOF spectra.
5. Select the matches within the error tolerance threshold (typically, ±200 ppm if internal calibration is done or ±400 ppm when relying on external calibration only), giving priority to the most abundant and basic protein detected in the shotgun experiment if multiple matches occur. In the example cited here above, the mean experimental *m/z* value of 9989.6 found in the whole-cell MALDI-TOF spectrum of *R. lacuscaerulensis* ITI-1157 bacterium [20] matches relatively well with the *m/z* theoretical value (9989.5) calculated for the monocharged ion arising from the 9988.5 Da polypeptide mentioned here above. In this case, the error tolerance is −10 ppm. No other possible match can be obtained with any of the other proteins detected by shotgun proteomics which are mentioned in Table 1. Therefore, the protein giving the *m/z* signal centered at 9989.6 is the DNA-binding protein HU, with its initial methionine removed and unacetylated.
6. Verify that the matches are corresponding to basic proteins and mainly are monocharged species on the MALDI-TOF mass spectra (*see* **Note 12**).

3.3.4 Selecting and Validating with a Representative Panel of Bacteria

1. Check the predicted occurrence of the identified proteins giving the highest MALDI-TOF *m/z* signals in the available genomes belonging to the considered taxon (*see* **Note 13**).
2. Extract the protein sequences of the most-related proteins.
3. Predict whether the initial methionine will be processed or not by cellular methionine aminopeptidases, and evaluate whether the observed acetylation is conserved throughout the taxon if any.
4. Select the most interesting biomarkers on the basis of their general occurrence throughout the taxon and their sequence similarities (*see* **Notes 14** and **15**).
5. Select representative strains belonging to the considered taxon for validating the potential biomarkers.
6. Record four MALDI-TOF spectra with four independent replicates of each representative strain.
7. Check directly on the spectra the presence of the expected *m/z* peaks or extract the 100 most intense *m/z* signals for processing the values with excel.
8. Keep the biomarkers that are consistently found in the MALDI-TOF spectra of the representative strains and exhibiting the most intense signals.

4 Notes

1. Bacteria used for establishing reference spectra and reference protein catalogs should be clonal, i.e., indistinguishable in genotype [24] and well characterized in terms of taxonomy and genomics [25]. As a general precaution, bacterial strains should be checked for purity on solid media prior to mass spectrometry. The genomes of several thousand strains have been sequenced and annotated but the quality of the resulting sequences and their annotation may not be always optimal [17, 26]. Choosing reference bacterial strains for establishing generic biomarkers should be done in the light of genome sequence availability and relevant annotation quality indices, such as number of resulting contigs, N50 value (which represents the average length of the final set of assembled sequences), and number of annotated genes compared to those of closely related strains.
2. Cells should be cultivated under optimal conditions on agar plates and harvested typically after 24 or 48 h growth depending on the organism to be considered. Long storage at optimal growth temperature or in the refrigerator at lower temperature should be avoided as the proteome will differ from the proteome in exponential growth phase. Measurements with MALDI-TOF and tandem mass spectrometry should be performed on cells harvested at the same time from the same plate.
3. Slow growing bacteria, typically small colonies forming on agar plates, may be directly cultured in liquid broth in optimized conditions of medium, shaking, oxygenation, and temperature. Process the pellet obtained after centrifugation of the liquid culture as a colony.
4. For BSL2 or BSL3 pathogens, samples must be manipulated under the appropriate safety conditions, inactivated prior to further processing, and the inactivation procedure should be checked for example by inoculation of an agar plate. For example, inactivate the bacterial biomass with 100 μL of 80% (vol/vol) aqueous trifluoroacetic acid per 10 mg of bacteria, incubate for 30 min at room temperature under gentle agitation (180 rpm), and then add 150 μL of 70% (vol/vol) acetonitrile and centrifuge at 16,000 × g for 5 min. Use the supernatant containing the soluble proteins as biological material for MALDI-TOF plate deposition and shotgun proteomics as described previously [23].
5. Enzyme solution: Reconstitute lyophilized sequencing-grade trypsin to a final concentration of 0.1 μg/μL in 0.01% trifluoroacetic acid. Reconstituted enzyme can be aliquoted and

stored at −20 °C for several months or at 4 °C for up to 1 week [27]. Passing reconstituted trypsin through multiple freeze–thaw cycles is not recommended.

6. In case of poor MALDI-TOF spectra, alternatives may be tried such as the addition of 70% formic acid before the addition of matrix [28]. Such alternative is recommended for Gram positive bacteria.
7. For optimal long-term storage, peptide solutions should be lyophilized in low-adsorption tubes which prevent peptides to stick to the tube walls.
8. MALDI-TOF mass spectrometry acquisition parameters may be optimized [29, 30]. For calibration of the MALDI-TOF mass spectrometer, the protein calibration standard I (Bruker Daltonics) can be used after appropriate dilution in 0.1% trifluoroacetic acid, the dilution being chosen to get a signal comparable in intensity to the sample peaks. This standard comprises four proteins giving rise to 6 *m/z* peaks: cytochrome C (*m/z* at 12360.97 and 6180.99 for $[M+H]^+$ and $[M+2H]^{2+}$, respectively), insulin (*m/z* at 5734.51 for $[M+H]^+$), myoglobin (*m/z* at 16952.30 and 8476.65 for $[M+H]^+$ and $[M+2H]^{2+}$, respectively), and ubiquitin I (*m/z* at 8565.76 for $[M+H]^+$).
9. For internal calibration, the protein calibration standard encompassing four to eight proteins should be added directly to the biological material just prior to deposition on the MALDI target plate.
10. The average molecular weight of a polypeptide can be calculated with its exact chemical formulae on the basis of normal distribution of all possible atom isotopes. A tool such as Compute pi/Mw (http://web.expasy.org/compute_pi/) can be used.
11. Posttranslational modifications of the proteins should be taken into account in order to evaluate the experimental average mass. The two most prevalent modifications in bacteria are initial methionine processing (−149.2 Da) by methionine aminopeptidases and N-terminal acetylation (+42.0 Da). While the first modification is universal, occurrence and extent of the second modification depends on the bacterium under scrutiny. Initial methionine removal occurs when the second residue of the polypeptide encompasses a small lateral chain [31]. Therefore, removal frequently occurs for glycine, alanine, serine, cysteine, threonine, proline, or valine as second residue in the polypeptide, but can result in some cases in only partial removal.
12. Proteins giving *m/z* values in the MALDI-TOF mass spectra acquired through the whole-cell MALDI-TOF approach are basic, abundant, and low molecular weight polypeptides. They are more often resulting from monocharged ion species and less

frequently from double charged ion species. Indeed, if a protein with a molecular weight below 20 kDa is really abundant, both forms may be observed, the single charged ion species being often more intense than the double charged ion species.

13. For identifying the most-closely related proteins in the available genomes belonging to the considered taxon, download the genome sequences of all the representative strains of the taxon. Identify the homologs of the putative biomarkers in each of these representative strains by automatic sequence alignment using the common basic local alignment search tool (BLAST).
14. Some biomarkers may be highly conserved, resulting in the same *m/z* values throughout the whole taxon, while some others will exhibit different *m/z* values due to some amino acid variations between closely related homologs, the latter being interesting to differentiate subgroups within the taxon.
15. Because genome annotation may not be always perfect, check any specific cases such as unexpected lower sequence similarity, absence of a conserved protein, or atypical N-terminus compared to the consensus sequences and reannotate by proteogenomics the corresponding protein sequences if required.

Acknowledgments

This work was supported by the Commissariat à l'Energie Atomique et aux Energies Alternatives, and the Agence Nationale de la Recherche (ANR-12-BSV6-0012-01). I thank my colleagues, Guylaine Miotello and Joseph Christie-Oleza, for stimulating discussions when elaborating this methodology.

References

1. Fenselau C, Demirev PA (2001) Characterization of intact microorganisms by MALDI mass spectrometry. Mass Spectrom Rev 20:157–171
2. Fenselau CC (2013) Rapid characterization of microorganisms by mass spectrometry—what can be learned and how? J Am Soc Mass Spectrom 24:1161–1166
3. Sauer S, Kliem M (2010) Mass spectrometry tools for the classification and identification of bacteria. Nat Rev Microbiol 8:74–82
4. Welker M (2011) Proteomics for routine identification of microorganisms. Proteomics 11:3143–3153
5. Chenau J, Fenaille F, Ezan E, Morel N, Lamourette P, Goossens PL, Becher F (2011) Sensitive detection of Bacillus anthracis spores by immunocapture and liquid chromatography-tandem mass spectrometry. Anal Chem 83:8675–8682
6. Charretier Y, Dauwalder O, Franceschi C, Degout-Charmette E, Zambardi G, Cecchini T, Bardet C, Lacoux X, Dufour P, Veron L et al (2015) Rapid bacterial identification, resistance, virulence and type profiling using selected reaction monitoring mass spectrometry. Sci Rep 5:13944
7. Duriez E, Armengaud J, Fenaille F, Ezan E (2016) Mass spectrometry for the detection of bioterrorism agents: from environmental to clinical applications. J Mass Spectrom 51:183–199

8. Karlsson R, Gonzales-Siles L, Boulund F, Svensson-Stadler L, Skovbjerg S, Karlsson A, Davidson M, Hulth S, Kristiansson E, Moore ER (2015) Proteotyping: proteomic characterization, classification and identification of microorganisms—a prospectus. Syst Appl Microbiol 38:246–257
9. Pribil P, Fenselau C (2005) Characterization of Enterobacteria using MALDI-TOF mass spectrometry. Anal Chem 77:6092–6095
10. Wynne C, Edwards NJ, Fenselau C (2010) Phyloproteomic classification of unsequenced organisms by top-down identification of bacterial proteins using capLC-MS/MS on an Orbitrap. Proteomics 10:3631–3643
11. Welker M, Moore ER (2011) Applications of whole-cell matrix-assisted laser-desorption/ionization time-of-flight mass spectrometry in systematic microbiology. Syst Appl Microbiol 34:2–11
12. Liu H, Du Z, Wang J, Yang R (2007) Universal sample preparation method for characterization of bacteria by matrix-assisted laser desorption ionization-time of flight mass spectrometry. Appl Environ Microbiol 73:1899–1907
13. Holland RD, Duffy CR, Rafii F, Sutherland JB, Heinze TM, Holder CL, Voorhees KJ, Lay JO Jr (1999) Identification of bacterial proteins observed in MALDI TOF mass spectra from whole cells. Anal Chem 71:3226–3230
14. Ryzhov V, Fenselau C (2001) Characterization of the protein subset desorbed by MALDI from whole bacterial cells. Anal Chem 73:746–750
15. Wynne C, Fenselau C, Demirev PA, Edwards N (2009) Top-down identification of protein biomarkers in bacteria with unsequenced genomes. Anal Chem 81:9633–9642
16. Armengaud J, Hartmann EM, Bland C (2013) Proteogenomics for environmental microbiology. Proteomics 13:2731–2742
17. Pible O, Hartmann EM, Imbert G, Armengaud J (2014) The importance of recognizing and reporting sequence database contamination for proteomics. EuPA Open Proteom 3:246–249
18. Fagerquist CK, Garbus BR, Miller WG, Williams KE, Yee E, Bates AH, Boyle S, Harden LA, Cooley MB, Mandrell RE (2010) Rapid identification of protein biomarkers of *Escherichia coli* O157:H7 by matrix-assisted laserdesorptionionization-time-of-flight-time--of-flight mass spectrometry and top-down proteomics. Anal Chem 82:2717–2725
19. Sun L, Teramoto K, Sato H, Torimura M, Tao H, Shintani T (2006) Characterization of ribosomal proteins as biomarkers for matrix-assisted laser desorption/ionization mass spectral identification of Lactobacillus plantarum. Rapid Commun Mass Spectrom 20:3789–3798
20. Christie-Oleza JA, Pina-Villalonga JM, Guerin P, Miotello G, Bosch R, Nogales B, Armengaud J (2013a) Shotgun nanoLC-MS/MS proteogenomics to document MALDI-TOF biomarkers for screening new members of the Ruegeria genus. Environ Microbiol 15:133–147
21. Christie-Oleza JA, Miotello G, Armengaud J (2013b) Proteogenomic definition of biomarkers for the large Roseobacter clade and application for a quick screening of new environmental isolates. J Proteome Res 12:5331–5339
22. Suarez S, Ferroni A, Lotz A, Jolley KA, Guerin P, Leto J, Dauphin B, Jamet A, Maiden MC, Nassif X et al (2013) Ribosomal proteins as biomarkers for bacterial identification by mass spectrometry in the clinical microbiology laboratory. J Microbiol Methods 94:390–396
23. Durighello E, Bellanger L, Ezan E, Armengaud J (2014) Proteogenomic biomarkers for identification of Francisella species and subspecies by matrix-assisted laser desorption ionization-time-of-flight mass spectrometry. Anal Chem 86:9394–9398
24. Spratt BG (2004) Exploring the concept of clonality in bacteria. Methods Mol Biol 266:323–352
25. Pible O, Armengaud J (2015) Improving the quality of genome, protein sequence, and taxonomy databases: a prerequisite for microbiome meta-omics 2.0. Proteomics 15:3418–3423
26. Armengaud J (2013) Microbiology and proteomics, getting the best of both worlds! Environ Microbiol 15:12–23
27. Hartmann EM, Allain F, Gaillard JC, Pible O, Armengaud J (2014) Taking the shortcut for high-throughput shotgun proteomic analysis of bacteria. Methods Mol Biol 1197:275–285
28. Schumann P, Maier T (2014) MALDI-TOF mass spectrometry applied to classification and identification of bacteria. Methods Microbiol 41:275–306
29. Lagace-Wiens P (2015) Matrix-assisted laser desorption/ionization time of flight mass spectrometry (MALDI-TOF/MS)-based identification of pathogens from positive blood culture bottles. Methods Mol Biol 1237:47–55
30. Petersen CE, Valentine NB, Wahl KL (2009) Characterization of microorganisms by MALDI mass spectrometry. Methods Mol Biol 492:367–379
31. Bonissone S, Gupta N, Romine M, Bradshaw RA, Pevzner PA (2013) N-terminal protein processing: a comparative proteogenomic analysis. Mol Cell Proteomics 12:14–28

Chapter 7

Detection and Typing of "*Candidatus* Phytoplasma" spp. in Host DNA Extracts Using Oligonucleotide-Coupled Fluorescent Microspheres

Edel Pérez-López, Christine Hammond, Chrystel Olivier, and Tim J. Dumonceaux

Abstract

The use of oligonucleotide-coupled fluorescent microspheres is a rapid, sequencing-independent, and reliable way to diagnose bacterial diseases. Previously described applications of oligonucleotide-coupled fluorescent microspheres for the detection and identification of bacteria in human clinical samples have been successfully adapted to detect and differentiate "Ca. Phytoplasma" species using as a target the *chaperonin 60*-encoding gene. In this chapter, we describe in detail the design and validation of oligonucleotide capture probes, and their application in the assay aiming to differentiate phytoplasma strains infecting *Brassica napus* and *Camelina sativa* plants grown in the same geographic location at the same time.

Key words Multiplex, Fluorescent microspheres, Phytoplasma, 16SrXIII group

1 Introduction

"*Candidatus* Phytoplasma" spp. are members of the class Mollicutes that infect plant and insect host species. While phytoplasma infection of phytophagous insects often results in increased host lifespan and fecundity [1], infection of plants as a result of feeding activity, and therefore phytoplasma inoculation, induces dramatic plant morphological changes in which floral parts become leaf-like [2], which greatly reduces seed set. More than 1000 plants including many crop and ornamental species have been identified worldwide as hosts [3], and hundreds of phloem-feeding insects, principally leafhoppers, have been identified as vectors that can transmit these phytopathogenic bacteria to new hosts [1]. Infection of crop plants can cause economically significant yield losses [3], so the detection and identification of phytoplasma infections in insect and plant hosts is of concern to the agricultural industry, in order to develop appropriate disease management strategies.

Kimberly A. Bishop-Lilly (ed.), *Diagnostic Bacteriology: Methods and Protocols*, Methods in Molecular Biology, vol. 1616, DOI 10.1007/978-1-4939-7037-7_7, © Springer Science+Business Media LLC 2017

Due to difficulty in establishing axenic cultures of phytoplasmas [4–6], the taxonomy of these bacteria is based on restriction fragment length polymorphism (RFLP) analysis of PCR amplified 16S rRNA gene sequences [2], supported by the phylogenetic analysis of the 16S rRNA gene. However, the use of other genes as part of the scheme of identification and classification of phytoplasmas has been broadly suggested. Protein-encoding genes typically show better strain resolution compared with ribosomal RNA-encoding genes [7]. DNA-dependent RNA polymerase β-subunit-encoding gene (*rpoB*), ribosomal protein *rplV–rpsC*-encoding genes, and the GroEL-encoding gene are some of the genes used recently to achieve a better differentiation of phytoplasma-related strains [8, 9].

Molecular diagnostic methods are used to detect, quantify, and identify infections of plant and insect hosts with phytoplasma. Polymerase chain reaction (PCR), real-time quantitative PCR, loop-mediated amplification isothermal (LAMP), and microarray have all been used successfully to detect phytoplasmas present in affected tissue, and the reported methods typically target 16S rRNA-encoding sequences [10–15].

We recently reported sets of PCR primers that can access the *cpn*60 universal target (*cpn*60 UT) of a diverse range of "*Ca.* Phytoplasma" spp. [16]. The *cpn*60 UT is a fragment of approximately 550 bp that has been defined as a molecular barcode for the domain Bacteria [17] and has been identified as a diagnostic target that provides improved phytoplasma strain resolution compared to 16S rRNA-encoding targets [16]. With *cpn*60 UT sequence data in hand, a range of molecular diagnostic assays can be developed to detect and differentiate "Ca. Phytoplasma" spp. in infected plant and insect hosts. Diagnostic methods that are rapid, sequencing-independent, and amenable to moderate- to high-throughput analysis are a valuable means of detection and identification of phytoplasma. We have described the application of oligonucleotide-coupled fluorescent microspheres for the detection and typing of bacteria in human clinical samples [18, 19], and have recently adapted that approach to the detection and differentiation of "Ca. Phytoplasma" spp. [16]. In the latter work, we developed an 11-plex assay capable of differentiating various groups of phytoplasmas that showed *cpn*60 UT sequence identities of ~65–97% [16]. This assay was capable of detecting simultaneous infections in plant hosts of two phytoplasma strains that showed a high sequence identity, and revealed a differential pattern of infection in two related crop species, *Brassica napus* and *Camelina sativa.*

In the present work, we describe these methods in the format of a laboratory protocol, with specific emphasis on capture oligonucleotide probe design for targeting the sequence of interest. The basic laboratory steps of the procedure are *cpn60* UT amplification using modified primers, coupling of capture oligonucleotides to fluorescent beads, and hybridization of the amplicon to the capture oligonucleotide-coupled microspheres. We previously described these steps in detail for the universal detection of bacterial strains, and the present work will focus on the application of the assay to the detection of "Ca. Phytoplasma" spp. in particular and will describe probe design and validation steps. To illustrate the probe design methods, we have provided details of probe design and validation from previously published probes targeting specific groups of phytoplasma [16]. We have applied the method described here to new phytoplasma-infected samples obtained from *B. napus* and *C. sativa* grown in the same field in Saskatoon, Saskatchewan, Canada, in the 2013 crop year, complementing our previously reported results [16] on these same plant species from the 2012 crop year [16]. The aim of this assay is to answer the diagnostic question, "Does this sample contain DNA from any of the phytoplasma strains represented on this array?" It offers a semiquantitative, rapid alternative to direct sequencing of amplicons, and in certain cases (for example, when simultaneous infections with two highly related strains of phytoplasma are present), it can provide improved resolution compared to direct sequencing.

2 Materials

2.1 Oligonucleotide Probe Design

1. Software to guide hybridization sites for oligonucleotides for maximum target specificity. We use sigoligo, a program that was initially designed for microarray work, to identify locations that will likely contain probes that are specific to the target of interest. Using sigoligo in combination with PrimerPlex v2.62 (Premier Biosoft, Palo Alto, CA) greatly facilitates the design of oligonucleotides that hybridize to the target of interest and not to amplicon generated from any of the other known *cpn60* UT sequences from "Ca. Phytoplasma" spp.

2.2 cpn60 Universal Target Amplicon Generation

1. Programmable thermocycler with a 96-well heating block and a heated lid (e.g., Bio-Rad C1000).
2. PCR primers recognizing the *cpn60* UT, 100, 25, or 10 μM working stocks. Upstream primers are modified at the 5′ end to contain biotin and phosphorothioate modified nucleotides (Table 1).
3. Thermostable DNA polymerase for PCR. The polymerase is chosen based on personal preference as well as empirical experience. A "hot start" polymerase is optional. We typically use Platinum Taq or Taq polymerase (Invitrogen).
4. 96-Well PCR plates or strip tubes compatible with the thermocycler.

Table 1
Sequences of the *cpn60* amplification primers

Primer name [16]	Primer sequence (5′–3′)[a]
H279-phyto-lum	5′-biotin-[*C][*G][*A][*C]GATIIIGCAGGIGATGGAACMACIAC
H280-phyto	TGRTTITCICCAAAACCAGGIGCATT
D0317-lum	5′-biotin-[*C][*G][*A][*C]GATIIIKCIGGIGAYGGIACIACIAC
D0318	TGRTKITCICCAAAACYWGGIGCWTC

Note that there are two upstream and two downstream primers that can be mixed to various molar ratios in order to amplify the *cpn60* UT from a diverse array of Phytoplasma spp. We have determined that a 1:7 molar ratio of H279p/H280p:D0317/D0317 successfully amplifies the *cpn60* UT from a wide range of "Ca. Phytoplasma" spp. [16]
[a]I = inosine; Y = C or T; R = A or G; K = T or G; S = C or G; [*] = phosphorothioate-modified nucleotide

5. T7 exonuclease (New England Biolabs).
6. 0.5 M ethylenediaminetetraacetic acid (EDTA) solution, pH 8.0.
7. Two workstations reserved for PCR work fitted with ultraviolet bulbs (e.g., PCR Cleanspot, Coy Laboratory Products). *See* **Note 1** for cautionary steps to be taken to avoid contamination.
8. Quant-iT DNA quantification kit (Invitrogen) or analogous fluorescence-based DNA quantification system and fluorometer (e.g., Qubit fluorometer, Invitrogen).

2.3 Coupling Oligonucleotides to Fluorescent Beads

1. EDC 1-ethyl-3-(3-dimethylamiopropyl) carbodiimide HCl.
2. 5.6 μm polystyrene Bio-Plex beads, various spectral signatures (colors). Magnetic beads may also be used.
3. Target-specific capture oligonucleotides, 5-amino-C12 modified (IDT, Invitrogen, Eurofins, or other suppliers).
4. 0.1 M MES pH 4.5 (protect this solution from light).
5. 0.02% Tween 20.
6. 0.1% sodium dodecyl sulfate.
7. TE buffer (10 mM Tris–Cl pH 8.0; 1 mM EDTA).

2.4 Hybridization of PCR Product to Oligonucleotide-Coupled Beads

1. Bio-Plex or analogous instrument (e.g., Luminex).
2. Streptavidin-R-phycoerythrin (SAPE), 1 mg/ml. High-purity SAPE solution is essential, e.g., Life Technologies cat no. S-866.
3. Thermowell 96-well PCR plates, low profile.
4. Thermowell silicone sealing mat. Can be reused.
5. 5 M tetramethyl ammonium chloride (TMAC), 10% sarkosyl, 1 M Tris–Cl, pH 8.0.

3 Methods

3.1 Oligonucleotide Probe Design

The objective of this analysis is to identify regions that are likely to contain sequences specific to the target sequence of interest. The output is a series of ranges that contain "signature" regions that are capable of discerning the target sequence from all of the other sequences in the "outgroup." Other strategies are equally valid but this is our typical approach. Capture oligonucleotide design for a previously described *cpn60* UT sequence from Bois Noir (strain BN45660; GenBank accession no. KJ939981) is used as an example.

1. Place the sequence files (in Fasta format) to be analyzed in an appropriately labeled folder (for the current example, this folder is labeled, "phyto").
2. Within this folder, arrange the sequences so that the sequence to be examined for signatures (in this case, BN-cpn60.fasta) is at the same level as another folder (here called, "outgroup") containing all of the sequences to be discerned from the target sequence. In this case, the "outgroup" folder contains *cpn60* sequences from all of the non-BN phytoplasmas that are known to date [16]. These can be retrieved from GenBank (ncbi.wnlm.nih.gov) or cpnDB (www.cpndb.ca).
3. Execute the sigoli command from a folder one level above the "phyto" folder, as follows:
   ```
   /path-to-sigoli-1-1/sigoli -operation = ranges -sequence-directory = phyto -oligo-size = 20 -diff=yes > BN-sigs.txt
   ```
   This specifies that the software should return the ranges and their mid-point (nucleotide positions) of signatures within the sequence directory, "phyto," with an oligonucleotide size of 20. The "diff=yes" flag specifies that the ranges it finds must differ in more than one place from all of the other sequences. The output of the program is directed to a new file called BN-sigs.txt.
4. Examine the output of sigoli; *see* Fig. 1a for an example. The output will show the ranges of nucleotide positions within BN-cpn60.fasta that contain at least two differences to all of the sequences in outgroup within a sliding window of 20 nucleotides. These areas are suitable for specific capture oligonucleotide design.
5. Enter all of the sequences into PrimerPlex. Execute a search for capture oligonucleotides (under Analyze > Capture Probe Search), ensuring that the "Design Anti-Sense Probe" radio button is selected. The number of capture probe sequences returned can be adjusted.

A

```
------------------------
b9979 DQ912815 Candidatus Phytoplasma sp. Crotalaria sp. Thailand' witches'-broom phytoplasma
------------------------
------------------------
b9975 DQ912819 Candidatus Phytoplasma solani StolC
------------------------
[   1 -  29] :  15
[  29 - 121] :  75
[ 125 - 166] : 145
[ 152 - 211] : 181
[ 194 - 282] : 238
[ 283 - 324] : 303
[ 323 - 391] : 357
[ 393 - 533] : 463
[ 524 - 545] : 534
[ 527 - 553] : 539
```

B

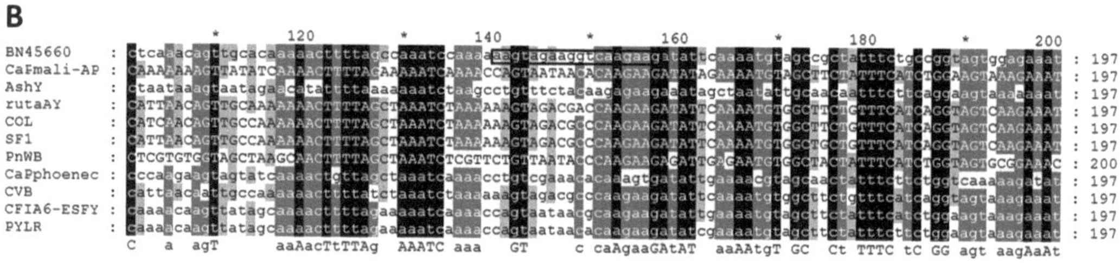

Fig. 1 An example of capture probe design. (**a**) Output of sigoligo when BN-cpn60.fasta was compared to *cpn60* sequence data representing ten other phytoplasma targets. (**b**) Location of designed probe for BN in the target sequence (BN45660) compared to all of the other Phytoplasma *cpn60* sequences considered. Note that the reverse complement of the probe sequence is indicated, since the probe hybridizes to the sense strand

6. Compare the locations of the designed capture probes with the signature locations specified in **step 3**. Select probes that land in identified signature regions for further analysis. The sequence of the oligonucleotide capture probe that was designed for BN is: 5′-amino-C12-CTTCTTGACCTTCTACTT-3′ (Fig. 1b).
7. The target specificity of the capture probes can be analyzed using primer BLAST at cpndb.ca or ncbi.nlm.nih.gov. *See* **Notes 2** and **3** for further details.
8. Sequences that share very high sequence identities may be discerned even if sigoligo returns few or no signature regions. To do this, the single nucleotide polymorphism (SNP) detection feature of PrimerPlex can be used. *See* Fig. 2a for an example. In the case of AY-COL, setting "C" at position 101 as the wild-type base and "T" as the mutant base and specifying SNP detection will facilitate the design of a probe that locates this difference near the middle of the sequence, which enhances target specificity (*see* **Note 3**). Figure 2b shows that nearly identical probes containing slight mismatches can still be highly specific to the target of interest. *See* Table 3, Fig. 2, and **Note 4** for discussion of an application of this resolving power.

3.2 cpn60 Amplicon Generation

Procedures to extract DNA to be used as a template for the *cpn60* UT amplification are not described. Normally, templates will consist of plant or insect hosts that are suspected to contain phytoplasma

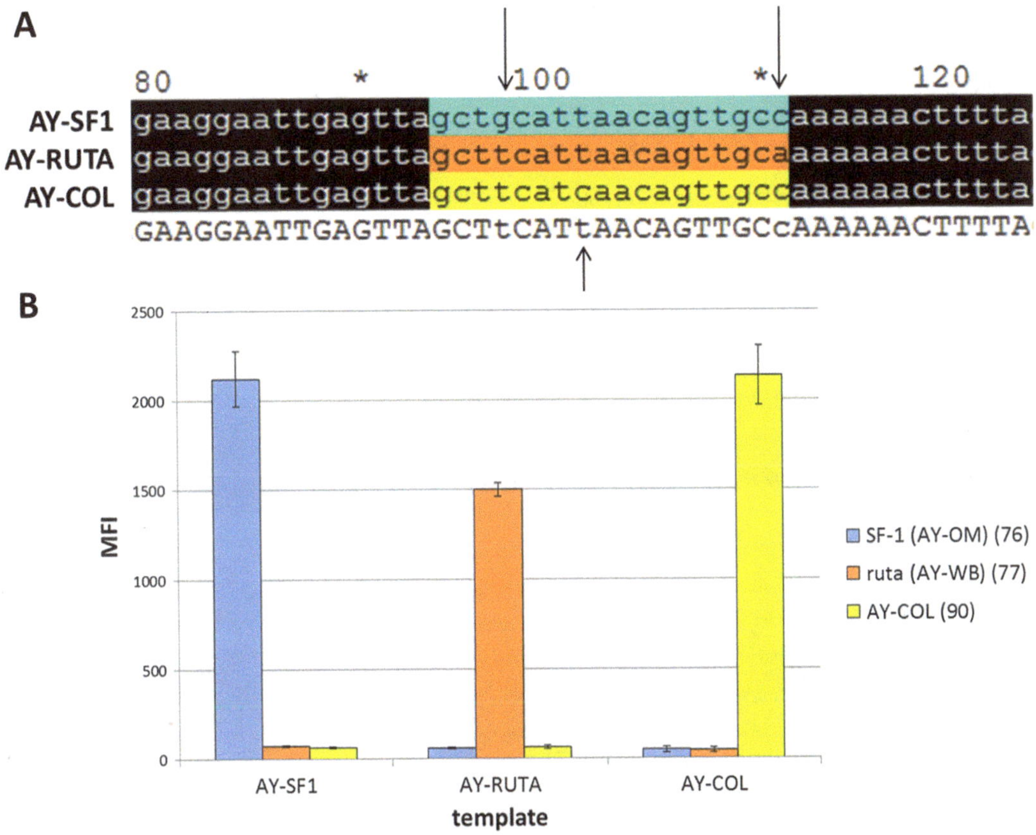

Fig. 2 Design of capture oligonucleotides capable of discerning sequences with high sequence identity. The sequences shown are 97–98% identical over their 552 bp lengths. (**a**) Alignment of three Aster Yellows sequences showing the portions of the sequences containing the capture probe hybridization sites. *Arrows* indicate the locations of differences that are exploited to generate target-specific capture oligonucleotides. (**b**) Results obtained with the designed oligonucleotides

DNA. For plant hosts, we typically use a commercial kit for DNA extraction. For insect hosts, we have used a modified version of the hexadecyltrimethylammonium bromide (CTAB)-based method described by Daire et al. [20], but without β-mercaptoethanol. Alternatively, we have used a commercial kit. For *Macrosteles*, we normally use six adults or ten nymphs per extraction, and for *Aceratagallia*, we use three adults or five to eight nymphs. The final volume of DNA extract in either case is 50 μl. Typically, this DNA is quantified using a Qubit or other fluorometric methods. As with any universal PCR method, it is best to use the "white-gray-black" system for handling mastermix, template, and amplicon—*see* **Note 1** for more details. The steps below describe what we have used for the generation of *cpn60* UT amplicon from "Ca. Phytoplasma" spp., which is distinct from the universal method for Bacteria [18].

Table 2
Suggested reaction setup for generating amplicon for fluorescent microsphere assay to detect "Ca. Phytoplasma" spp.

Mastermix 1: buffer, Mg, and dNTP (prepare ahead of time and store at −20 °C)				
Component	μl/reaction	μl/100 reactions	μl/300 reactions	Final concentration in PCR
10× PCR buffer	5	500	1500	1×
50 mM $MgCl_2$	2.5	250	750	2.5 mM
10 mM dNTP	2.5	250	750	0.5 mM (each)
H279p-lum, 100 μM	0.025	2.5	7.5	50 nM
H280p, 100 μM	0.025	2.5	7.5	50 nM
D0317-lum, 100 μM	0.175	17.5	52.5	350 nM
D0318, 100 μM	0.175	17.5	52.5	350 nM
Water	37.4	3740	11,220	–
Total	47.8	4780	14,340	–
Mastermix 2: Mixture for distribution into plates/tubes and template addition				
Component	μl/reaction			
Mmx1	47.8			
Taq DNA polymerase, 5 U/μl	0.2			
Total	48			

This setup uses a 1:7 molar ratio of H279-phyto/H289-phyto:D0317:D0318, but other ratios can be used. We typically use a final primer concentration of 400 nM each of the upstream and downstream primers

1. Prepare a PCR mastermix sufficient for the desired number of reactions. Set up the mastermix according to Table 2. This is a suggested setup for *Taq* DNA polymerase; for other polymerases and/or other mastermix volumes the values may need to be adjusted. Dispense this mastermix into aliquots and store at −20 °C.
2. Distribute 48 μl of this mastermix into each PCR plate well or strip tube. Generate two PCRs for each template to be analyzed. There should be sufficient mastermix to prepare two "no template" controls.
3. Add 2 μl template DNA to the reactions as required. A higher template volume can be added, but the volume of water added

to the mastermix must be adjusted accordingly. Add 2 μl of PCR-grade water to the no template controls.

4. Seal the plate with an appropriate PCR lid (or cap the tubes), and then place it in the thermocycler. Cycle the reactions under the following conditions:

   95 °C 5 min (1×).

   95 °C 30 s; 42 °C 30 s; 72 °C 30 s (40×).

   72 °C 2 min (1×).

5. Analysis of the reaction products by agarose gel electrophoresis is not necessary but can be done using 5–10 μl if desired.
6. Add 2 μl of T7 exonuclease to each reaction, mix, and incubate at room temperature for 40 min to 2 h. Stop the reaction by adding 12.5 μl of 500 mM EDTA. These tubes contain single-stranded amplicon that is ready for hybridization to the oligonucleotide-coupled fluorescent bead mixture.

3.3 Coupling Oligonucleotides to Fluorescent Beads and Hybridization of PCR Product to Oligonucleotide-Coupled Beads

The bead coupling method has been described [18] but is elaborated again here for completeness.

1. Before starting, decide which bead "color" (specified by a number on the package) will be coupled with each capture oligonucleotide. Keep careful track of this, as the instrument will report its results by the bead color and correct identification depends on the correspondence between the coupled oligonucleotide and the bead color.
2. EDC powder should be stored desiccated at −20 °C. Place EDC powder on the bench to warm up for a few minutes before starting.
3. Vortex the stock solution of the fluorescent beads to resuspend them.
4. Transfer 400 μl of microspheres to an Eppendorf tube. Alternatively, 100 μl of beads can be coupled for a smaller-scale preparation. This is useful if a new capture probe is being evaluated to avoid using up a large volume of bead solution. In this case, volumes are adjusted in subsequent steps, indicated in parentheses.
5. Pellet the microspheres at 14,000 × *g* for 1 min. Remove and discard the supernatant.
6. Resuspend the pellet in 50 μl (12.5 μl) of 0.1 M MES pH 4.5.
7. Prepare a fresh solution of EDC at 10 mg/ml in water. Prepare less than 1 ml.
8. Add 1 nmol capture oligonucleotide to the microspheres and mix by vortexing. For a 100 μM stock probe concentration, add 10 μl for 1 nmol.

9. Add 2.5 μl of fresh EDC solution to microspheres.
10. Incubate at room temperature 30 min in the dark.
11. Discard EDC solution from **step 7** and prepare a fresh sample of 10 mg/ml EDC in water.
12. Add another 2.5 μl of fresh EDC solution to the microspheres—vortex 10 s to mix well.
13. Incubate at room temperature for 30 min in the dark.
14. Add 1 ml of 0.02% Tween 20 and vortex to wash.
15. Centrifuge 14,000 × *g* 1 min. Remove and discard the supernatant.
16. Add 1 ml of 0.1% SDS and vortex to wash.
17. Centrifuge 14,000 × *g* 1 min. Remove and discard the supernatant.
18. Suspend beads in 100 μl (25 μl) of TE buffer.
19. Enumerate beads in haemocytometer or Coulter counter to determine the concentration of each bead suspension.
20. Store stock solutions of coupled beads at 4 °C in the dark. Beads may exhibit increased aggregation after months of storage in these conditions, so it is best to prepare the diluted bead mastermix right away.
21. Prepare a bead mastermix by mixing all of the various capture oligonucleotide-coupled fluorescent beads to a final concentration of 100 beads/μl each using TE buffer as a diluent. Store the mastermix at 4 °C in the dark. The mastermix is stable for many months under these conditions.

3.4 Hybridization of PCR Product to Oligonucleotide-Coupled Beads

1. While the PCR is running, perform Bio-Plex instrument warm-up and calibration steps according to the manufacturer's recommendations. The instrument should be set to a block temperature of 60 °C and should be at this temperature by the end of the hybridization step. Ensure that the needle is set to the appropriate height for PCR product analysis, according to the manufacturer's instructions.
2. Vortex or pipette to mix the stored fluorescent bead mastermix prepared in Subheading3.3. Determine the volume required for the experiment by considering that 33 μl will be added to each hybridization reaction. Dispense an appropriate amount of resuspended bead mastermix into a separate tube and sonicate in a waterbath sonicator for 5 min. Store the bead mastermix on ice after sonication.
3. Mix 17 μl of PCR product after T7 exonuclease treatment (*see* Subheading 3.2, **step 6**) with 33 μl of the sonicated fluorescent bead mastermix. Use a Thermowell low-profile or analo-

gous PCR plate and cover with a reusable, removable silicone cover.

4. Prepare fresh streptavidin-phycoerythrin (SAPE) solution sufficient for adding 25 µl per hybridization reaction. Dilute stock solution (1 mg/ml) 1:50 into tetramethylammonium chloride (TMAC) buffer: 3 M TMAC, 0.1% sarosyl, 50 mM Tris–HCl pH 8.0, 4 mM EDTA.
5. Incubate the PCR plate in an appropriate thermocycler with heated lid using the following program: 95 °C, 5 min; 60 °C, 10 min; 60 °C pause; 60 °C, 5, min.
6. At the 60 °C pause step, open the thermocycler, leaving the plate at 60 °C. Remove the silicone cover and add 25 µl of diluted SAPE solution to each hybridization reaction. Replace the cover, close the lid of the thermocycler, and resume the program.
7. When the program is complete, remove the plate from the thermocycler and quickly transfer it to the Bio-Plex instrument. The plate should not be allowed to cool significantly or nonspecific hybridization may occur. Be sure to remove the silicone cover, as the needle will not puncture it. Read the plate using the Bio-Plex software interface, specifying that the MFI be determined by reading 100 beads per region. Gate settings may need to be adjusted according to the beads used (magnetic or polystyrene).
8. Analyze the data using either an arbitrary positivity cutoff of 100–150 MFI or by comparing the MFI of the negative control (no template) to that of the sample. *See* **Note 5** for further discussion.

3.5 Assessment of Capture Probe Target Specificity

1. Set up a series of amplifications using samples from each of the different phytoplasma groups. Normally, we use plasmid DNA consisting of the *cpn60* UT of each phytoplasma sample, diluted to 5×10^6 copies per µl (*see* **Note 6**). When validating a new probe, it is often germane to include samples of genomic DNA extracted from tissues known to be infected with the target strain of interest. This will demonstrate that the probe is capable of detecting an infection with the target strain in a relevant sample, as opposed to the artificial situation of plasmid DNA in the solution. Be sure to include "no template" controls.
2. Prepare a microsphere mastermix containing capture oligonucleotide-coupled fluorescent microspheres detecting all of the phytoplasma strains so far described [16].
3. Hybridize amplicon to the multiplex microsphere mastermix as described in Subheading 3.3.

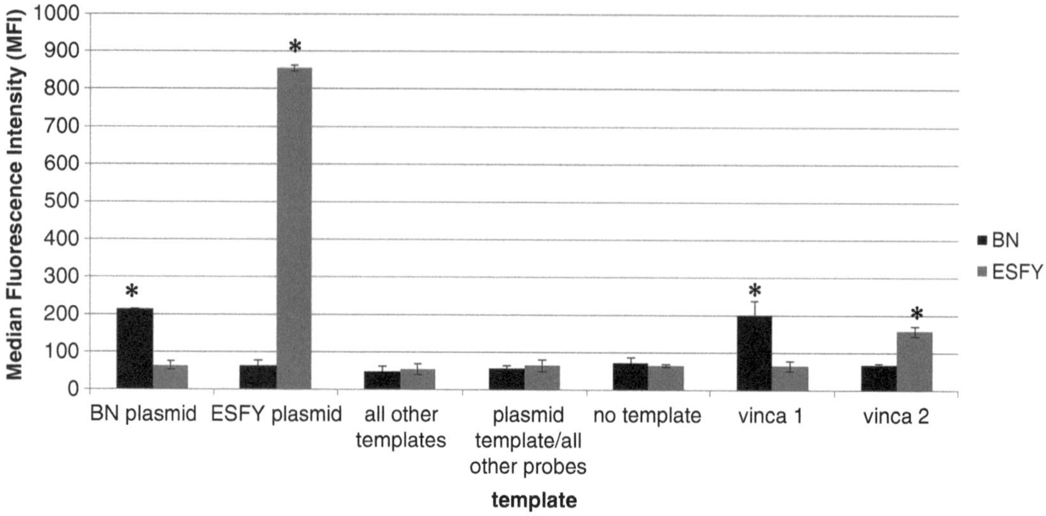

Fig. 3 Assessment of the performance of probes designed to detect Bois Noir (BN) and European Stone Fruit Yellows (ESFY) phytoplasmas. Samples with significantly positive MFI compared to no template controls are indicated as *asterisk*. Results generated using genomic DNA samples extracted from *Vinca* sp. (European periwinkle) infected with BN or ESFY phytoplasma are shown along with controls. Data is taken from Dumonceaux et al. [16]

4. Examine the MFI data: There should be a strong signal in the samples containing the target sequence of interest but no significant difference in MFI between the "no template" control and all of the other templates. *See* Fig. 3 and **Note 7** for further discussion.

4 Notes

1. PCR mastermix is typically prepared in a PCR workstation. We use separate workstations for mastermix preparation and template addition; each workstation has a set of dedicated pipettors and is regularly treated with ultraviolet radiation (the workstations have integrated bulbs that should be changed every year or two). Aerosol-resistant pipette tips are also essential for PCR setup. The pipettors and workstation should be regularly decontaminated using a product designed to remove DNA to minimize contamination. The mastermix preparation workstation ("white") should be kept free of template DNA and amplicon, and should be in a separate location from the template addition workstation ("gray"). Any handling of open tubes containing PCR products should be done in another separate area ("black"), and the workflow should be arranged such that technicians proceed only from white to gray to black areas on a given day.

2. In some cases, it is useful to analyze alignments of the target sequences and designed capture probes to best judge target specificity. This process requires a great deal of trial and error and patience, and in the end probes must be assessed for their performance in vitro rather than in silico. Although probes are not expensive to purchase, there is time and effort involved in the assessment of their performance, so care should be taken at the in silico analysis step to design a probe that is likely to perform in the desired manner (i.e., will generate a signal with the desired target but not with any of the other possible targets).
3. To maximize the chances of finding a target-specific oligonucleotide, look for differences near the middle of the designed probe. This is different from PCR primers, where target specificity is often obtained by setting the differences at the 3′ end of the primer. Hybridization probes that contain mismatches near the middle of the sequence are more destabilizing compared to differences at the ends of the probe sequences, and this can be exploited to discern sequences that are very highly related (Fig. 2).
4. In our previous work, we showed that the application of this assay to plant DNA extracts from infected plants grown in the same geographic location at the same time revealed a difference in host plant susceptibility wherein *C. sativa* showed no evidence of infection with AY strain ruta while *B. napus* showed evidence of both AY strain ruta and strain SF-1, and sometimes both strains. This difference was discernable using the probes shown in Fig. 2, which hybridize to the same location in the *cpn60* gene but contain two sequence differences relative to one another. Despite these small differences in capture probe sequence, virtually no nonspecific signal was detected (Fig. 2). To determine if this apparent difference in host plant susceptibility is maintained across growing years, we applied this assay to 50 *B. napus* and 49 *C. sativa* DNA extracts from infected plants grown in the same field during the 2013 growing season. As shown in Table 3, some of the *C. sativa* samples

Table 3
Number of positive samples determined using the oligonucleotide-coupled fluorescent microsphere assay on DNA samples extracted from *Brassica napus* ($n = 115$) or *Camelina sativa* ($n = 119$)

	2012[a]		2013		Proportion (total—2 year)	
Host	AY-ruta	AY-SF1	AY-ruta	AY-SF1	AY-ruta	AY-SF1
Brassica napus	23	42	17	33	34.78%	65.22%
Camelina sativa	0	70	7	42	5.88%	94.12%

[a]2012 data is taken from Dumonceaux et al. [16]

showed evidence of infection with AY-ruta (compared to none in the 2012 growing season), but the proportion of AY-ruta-positive plants was much lower in *C. sativa* (5.88%) compared to *B. napus* (34.78%). Note that this subtle difference would have been very difficult to detect using direct sequencing of amplicons because there are only 11 points of difference in the 552-bp *cpn60* UT sequences of these two strains. The biological significance of this observation is not presently known.

5. Because this is a semiquantitative method that generates numerical data in the form of MFI, there is some flexibility in terms of defining a positive result. For example, the Bio-Plex software allows the user to specify a cutoff MFI above which a sample is defined as positive for a given region. This may be suitable in many cases, but the "no template" control should be carefully examined as sometimes the MFI in the negative control can be quite variable. Another option is to use a statistical test such as a Student's *t*-test to determine if the MFI of a sample is significantly greater than the negative control at a given level (e.g., 0.01 or 0.05) [18, 19]. In any event, it is best to examine at least duplicate amplifications of a given template; if this is done there is sufficient volume of amplicon to examine up to four replicates from a sample. Normally, we use two replicates derived from two amplifications (i.e., one hybridization per amplification reaction) to determine positivity.
6. To prepare standards for this analysis, we typically use miniprep DNA from *Escherichia coli* cultures with plasmids containing the *cpn60* UT of all of the phytoplasma represented in the array. The plasmids are created by cloning the PCR products into a typical cloning vector such as pGEM-T Easy (Promega). DNA concentration is measured after miniprep using a Qubit or nanodrop instrument and the ng/μl results are converted to copies/μl using a molecular weight of 650 g/mol per base pair. The total length of the plasmid, including the backbone and *cpn60* UT insert (604 bp including primer sequences), must be considered in this calculation. For example, for pGEM-T Easy (3015 bp), the molecular weight of the plasmid can be estimated as $(3015 + 604) \times 650 = 2{,}352{,}350$ g/mol.
7. Despite all of the care that is taken during capture oligonucleotide design, the only way to be certain that a probe is specific for the target of interest is to examine its performance in vitro. An example of this for the BN and ESFY probes is shown in Fig. 3. Significant fluorescence signals were observed using templates consisting of the cloned BN or ESFY *cpn60* plasmid DNA. In contrast, no significantly positive MFI signals were observed when the BN or ESFY probes were used to query amplicon generated using all of the nonspecific templates, and the average MFI readings for all other probes with amplicons from the BN or ESFY *cpn60* plasmids were not significant

compared to the no template control. Application of these probes using genomic DNA taken from different *Vinca* sp. plants both detected the presence of phytoplasma DNA and determined the type of phytoplasma (BN or ESFY) without the need for DNA sequencing.

References

1. Weintraub PG, Beanland L (2006) Insect vectors of phytoplasmas. Annu Rev Entomol 51:91–111
2. Maejima K, Oshima K, Namba S (2014) Exploring the phytoplasmas, plant pathogenic bacteria. J Gen Plant Pathol 80(3):210–221. doi:10.1007/s10327-014-0512-8
3. Gasparich GE (2010) Spiroplasmas and phytoplasmas: microbes associated with plant hosts. Biologicals 38(2):193–203
4. Contaldo N, Bertaccini A, Paltrinieri S, Windsor HM, David Windsor G (2012) Axenic culture of plant pathogenic phytoplasmas. Phytopathol Mediterr 51(3):607–617
5. Zhao Y, Davis RE, Wei W, Shao J, Jomantiene R (2014) Phytoplasma genomes: evolution through mutually complementary mechanisms, gene loss and horizontal acquisition. In: Gross DC (ed) Genomics of plant-associated bacteria. Springer-Verlag, Berlin, Heidelberg, pp 235–271. doi:10.1007/978-3-642-55378-3_10
6. Zhao Y, Davis RE, Wei W, Lee IM (2015) Should 'Candidatus Phytoplasma' be retained within the order Acholeplasmatales? Int J Syst Evol Microbiol 65(Pt 3):1075–1082. doi:10.1099/ijs.0.000050
7. Zeigler DR (2003) Gene sequences useful for predicting relatedness of whole genomes in bacteria. Int J Syst Evol Microbiol 53(Pt 6):1893–1900
8. Valiunas D, Jomantiene R, Davis RE (2013) Evaluation of the DNA-dependent RNA polymerase β-subunit gene (*rpoB*) for phytoplasma classification and phylogeny. Int J Syst Evol Microbiol 63(Part 10):3904–3914
9. Mitrović J, Kakizawa S, Duduk B, Oshima K, Namba S, Bertaccini A (2011) The *groEL* gene as an additional marker for finer differentiation of '*Candidatus* Phytoplasma asteris'-related strains. Ann Appl Biol 159(1):41–48. doi:10.1111/j.1744-7348.2011.00472.x
10. Smart CD, Schneider B, Blomquist CL, Guerra LJ, Harrison NA, Ahrens U, Lorenz KH, Seemuller E, Kirkpatrick BC (1996) Phytoplasma-specific PCR primers based on sequences of the 16S-23S rRNA spacer region. Appl Environ Microbiol 62(8):2988–2993
11. Lee IM, Gundersen-Rindal DE, Davis RE, Bartoszyk IM (1998) Revised classification scheme of phytoplasmas based on RFLP analyses of 16S rRNA and ribosomal protein gene sequences. Int J Syst Bacteriol 48(4): 1153–1169
12. Obura E, Masiga D, Wachira F, Gurja B, Khan ZR (2011) Detection of phytoplasma by loop-mediated isothermal amplification of DNA (LAMP). J Microbiol Methods 84(2):312–316. doi:10.1016/j.mimet.2010.12.011
13. Hodgetts J, Tomlinson J, Boonham N, González-Martín I, Nikolić P, Swarbrick P, Yankey EN, Dickinson M (2011) Development of rapid in-field loop-mediated isothermal amplification (LAMP) assays for phytoplasmas. Bull Insectol 64(Suppl 1):S41–S42
14. Kogovšek P, Hodgetts J, Hall J, Prezelj N, Nikolić P, Mehle N, Lenarčič R, Rotter A, Dickinson M, Boonham N, Dermastia M, Ravnikar M (2015) LAMP assay and rapid sample preparation method for on-site detection of flavescence dorée phytoplasma in grapevine. Plant Pathol 64(2):286–296. doi:10.1111/ppa.12266
15. Jawhari M, Abrahamian P, Sater AA, Sobh H, Tawidian P, Abou-Jawdah Y (2015) Specific PCR and real-time PCR assays for detection and quantitation of 'Candidatus Phytoplasma phoenicium'. Mol Cell Probes 29(1):63–70. doi:10.1016/j.mcp.2014.12.003
16. Dumonceaux TJ, Green M, Hammond C, Perez E, Olivier C (2014) Molecular diagnostic tools for detection and differentiation of Phytoplasmas based on chaperonin-60 reveal differences in host plant infection patterns. PLoS One 9(12):e116039. doi:10.1371/journal.pone.0116039
17. Links MG, Dumonceaux TJ, Hemmingsen SM, Hill JE (2012) The chaperonin-60 universal target is a barcode for bacteria that enables *de novo* assembly of metagenomic sequence data. PLoS One 7(11):e49755. doi:10.1371/journal.pone.0049755

18. Dumonceaux TJ, Town JR, Hill JE, Chaban BL, Hemmingsen SM (2011) Multiplex detection of bacteria in complex clinical and environmental samples using oligonucleotide-coupled fluorescent microspheres. J Vis Exp:e3344. doi:10.3791/3344
19. Dumonceaux TJ, Schellenberg J, Goleski V, Hill JE, Jaoko W, Kimani J, Money D, Ball TB, Plummer FA, Severini A (2009) Multiplex detection of bacteria associated with normal microbiota and with bacterial vaginosis in vaginal swabs by use of oligonucleotide-coupled fluorescent microspheres. J Clin Microbiol 47(12): 4067–4077. doi:10.1128/jcm.00112-09
20. Daire X, Clair D, Reinert W, Boudon-Padieu E (1997) Detection and differentiation of grapevine yellows phytoplasmas belonging to the elm yellows group and to the stolbur subgroup by PCR amplification of non-ribosomal DNA. Eur J Plant Pathol 103(6):517–514

Chapter 8

Detection of *Helicobacter pylori* in the Gastric Mucosa by Fluorescence In Vivo Hybridization

Silvia Fontenete, Marina Leite, Ceu Figueiredo, Paul Cos, and Nuno F. Azevedo

Abstract

In this chapter, we describe a fluorescence in vivo hybridization (FIVH) protocol, using nucleic acid probes, for the detection of the bacterium *Helicobacter pylori* in the gastric mucosa of an infected C57BL/6 mouse model. This protocol should be easily extended to other microorganisms not only as a way to identify in vivo important microorganisms and their patterns of distribution within specific or at different anatomic sites, but also to better understand interaction mechanisms involving the microbiome and the human body.

Key words Diagnostics, Microbiology, Nucleic acids, In vivo, *Helicobacter pylori*, Fish, FIVH

1 Introduction

Since it was first developed, fluorescence in situ hybridization (FISH) has become one of the most frequently used molecular techniques in the microbiology field [1]. FISH is a powerful molecular method with widespread use in environmental and in clinical applications for the identification, visualization, and quantification of organisms of interest present in microbial communities [1–6]. Different FISH assays have been recently developed for the direct identification of a wide range of Gram-positive and Gram-negative bacteria in clinical samples [7, 8].

FISH is based on the annealing of fluorescently labeled oligonucleotides (commonly called probes) to a specific complementary target sequence, enabling its detection and quantification. When FISH is used as an identification method in bacteria, the target sequences are mainly selected within the 16S or 23S ribosomal RNA (rRNA), since these regions can be used as phylogenetic

Kimberly A. Bishop-Lilly (ed.), *Diagnostic Bacteriology: Methods and Protocols*, Methods in Molecular Biology, vol. 1616, DOI 10.1007/978-1-4939-7037-7_8, © Springer Science+Business Media LLC 2017

markers and rRNA is also present in a larger number of copies. When the sample is exposed to light of specific wavelengths, fluorescence can be detected by epifluorescence microscopy.

The design of adequate probes is a crucial step, and a previous and accurate analysis through available databases is required to achieve good levels of specificity and sensitivity [9–11]. Probes are synthesized and usually coupled at the 5′-end of the oligonucleotide with a fluorescent label dye (e.g., cyanine (Cy): Cy3, Cy5, Fluorescein amidite (FAM), Alexa fluor® dyes) [12, 13]. FISH protocols in bacteria usually comprise three steps: fixation/permeabilization, hybridization, and washing (Fig. 1). Fixation and permeabilization are typically joined in one operation with the objective to render the cell wall permeable to the nucleic acid probe while, at the same time, avoiding cell lysis and extensive nucleic acid degradation. During hybridization, the probe is placed in contact with the target cells, and if complementary (or near-complementary) sequences are present, hybridization will take place. The specificity of this binding event, i.e., the ability of the method to discriminate the target organisms from the remaining cells, is further ensured by a washing step where all loosely bound probes are washed away. Specificity can be theoretically predicted using 16S rRNA comparative sequence analysis, as probes are designed to confer a required level of taxonomic specificity (e.g.,

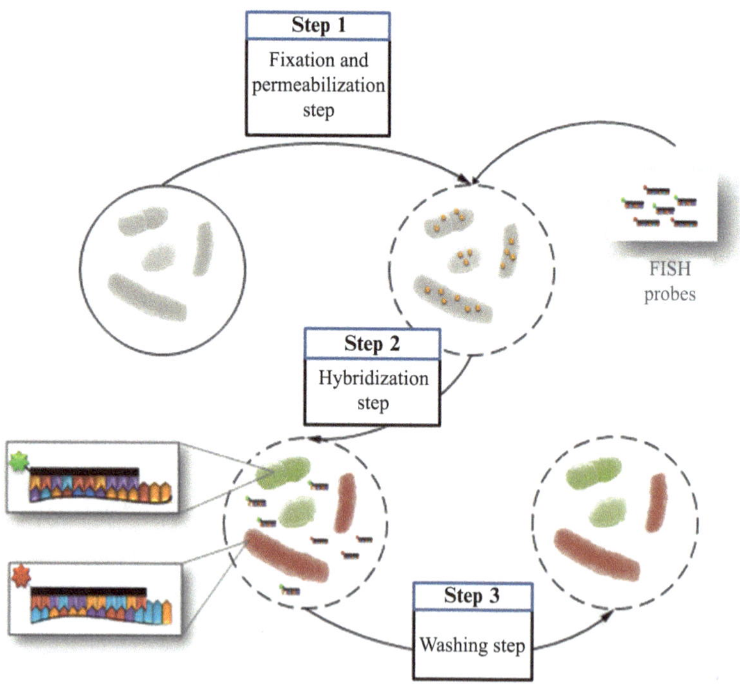

Fig. 1 Basic steps of fluorescence in situ hybridization in bacteria

species, genus, class), and then implemented in the laboratory by optimizing the hybridization and washing conditions [14]. Finally, visualization by either fluorescence microscopy or flow cytometry allows one to observe if successful hybridization has occurred.

The performance of fluorescence in vivo hybridization (FIVH) directly in the gastric mucosa implies that the full process has to be carried out at 37 °C. Because DNA oligonucleotides are not sufficiently reliable and versatile to be used in robust FISH methodologies, new synthetic monomers comprising novel chemical modifications with stronger target affinities can be used [15]. Locked nucleic acid (LNA) probes have been used for the detection of bacteria using FISH [16, 17]. However, the introduction of interspersed 2′ O-Methyl RNA (2′OMe) monomers and backbone modifications (phosphorothioate (PS) linkages instead of standard phosphodiester (PO)) showed high efficiency in *H. pylori* detection in vitro and in vivo [18–20].

This chapter is focusing on the in vivo detection of *H. pylori*, a well-known risk factor for gastric inflammation, peptic ulcer disease, and cancer in humans [21–23], using the FIVH methodology with LNA/2′OMe probes, specifically designed to target a *H. pylori* gene.

2 Materials

2.1 Probe Design

1. Ribosomal Database Project II (RDP-II) (available from https://rdp.cme.msu.edu/) software for probe selection.
2. BLAST (available from http://blast.ncbi.nlm.nih.gov/Blast.cgi) software for in silico alignment of sequences.

2.2 Probe Synthesis and Administration

1. Automated DNA synthesizer (PerSpective Biosystems Expedite 8909 instrument).
2. Universal polystyrene-based support.
3. Detritylation reagent: trichloroacetic acid in diclorometano (3:97, v/v).
4. Activator: 0.25 M of 4,5-dicyanoimidazole (DCI) in acetonitrile.
5. Cap A solution: acetic anhydride in tetrahydrofuran (THF) (9:91, v/v).
6. Cap B solution: *N*-methylimidazole in THF (1:9, v/v).
7. Thiolation solution: 0.0225 M xanthan hydrate in pyridine/acetonitrile (20:90, v/v).
8. Cleavage solution: 98% aqueous methanol/ammonia solution 7 N in methanol (1:1).

9. Cyanine phosphoramidite, Cy3 (Glen Research, VA, USA).
10. Phase high-performance liquid chromatography (HPLC) (RP-HPLC) using a Waters 600 system equipped with an XBridge OST C18 (2.5 μm, 19 × 100 mm) column and an XBridge Prep C18 (5 μm, 10 × 10 mm) precolumn.
11. Dionex system HPLC (VWR).
12. Matrix-assisted laser desorption ionization time-of-flight mass spectrometry (Bruker instruments, Leipzig, Germany).

2.3 Probe Synthesis and Administration

1. Female C57BL/6JRj strain mice (Janvier LABS Le Genest-St-Isle, France).
2. Adjuvant buffer: Prepare a buffer containing 0.5 M urea and 900 mM NaCl in water.
3. Polyethylene catheters (Biotrol, Paris, France).

2.4 FIVH Analysis

1. Paraformaldehyde solution: 4% paraformaldehyde (Sigma-Aldrich) in PBS.
2. Optimal cutting temperature compound: Tissue-Tek* O.C.T. Compound (Sakura FInetek, USA).
3. Cryostat.
4. Microtome.
5. Confocal system: Nikon Eclipse Ti-E inverted microscope attached to a microlens-enhanced dual spinning disk confocal system (UltraVIEW VoX; PerkinElmer, Seer Green, UK).

3 Methods

Carry out all procedures at room temperature, unless otherwise specified. The fluorescence probe solutions should be handled so that they are protected from the light to prevent degradation.

3.1 Probe Design

The probe selected in this study targets the 16S rRNA sequence of *H. pylori* (Table 1) [2, 24–26] (*see* **Note 1**). The use of nucleic acid mimics such as LNA and 2′OMe substitutions allows an increase in the duplex stability comparatively to DNA and RNA probes [27, 28]. This factor is crucial for the use of probes in FISH at low temperatures. However, some studies should be performed to analyze if the designed probe can be used with high efficiency, such as:

1. Analyze the thermodynamic parameters of the selected probe, for example free Gibbs energy [29].
2. Perform an in silico study to evaluate if the probe has a high specificity and sensitivity. Compare the probe sequence against sequences of *H. pylori* and closely related microorganisms using the freely available 16S rRNA database of the Ribosomal

Table 1
Probe sequenced used in this study for *H. pylori* detection

Designation	Sequence
Cy3 HP_LNA/2OMe_PS	5′-Cy3 G^L**ACT**^{L}**AAG**^{L}**CCC**^{L}-3′

Abbreviations: *LNA* L superscript, *2′-OMe* in boldface, *Cyanine* Cy3 Cy3_HP_LNA/2OMe_PS is a phosphorothioate oligomer (PS backbones)

Database Project II (RDP-II), version 10 [30]. Select only high-quality sequences with ≥1200 bp. The selected probe should differ by at least two mismatches from non-*H. pylori* species. Analyze the specificity of the probe against the host organism in which the infection will be performed, e.g., mouse, human. This study can be performed using BLAST.

3.2 Fluorescence in Vivo Hybridization (FIVH) Probe Synthesis

3.2.1 Synthesis

Synthesize LNA/2′OMe probe at 1.0 μmol scale using standard phosphoramidite chemistry and synthesis conditions described in Subheading 2, with an automated DNA synthesizer in a universal polystyrene-based support at 1.0 μmol scale, under anhydrous conditions:

1. Set the coupling time to 4.6 min for both monomers (*see* **Note 2**).
2. Add Cy3 in anhydrous acetonitrile at 0.1 M and activate it using tetrazole with a 20 min of coupling time (*see* **Note 3**).
3. Analyze the stepwise coupling yields by the absorbance of the dimethoxytrityl cations (DMT^+) released after each coupling step (*see* **Note 4**).
4. Cleave the probe from the support using the cleaving solution 2 h at room temperature followed by 32% v/v aqueous ammonia solution, 12 h at 55 °C.

3.2.2 Purification and Characterization of FIVH Probe

1. Purify the probe by reverse RP-HPLC.
2. After purification, precipitate the probe in 100% (v/v) acetone (Brand).
3. Characterize the purity and composition of the probe by IonExchange HPLC conditions (IE-HPLC) using a Dionex system HPLC (VWR) and matrix-assisted laser desorption ionization time-of-flight mass spectrometry (MALDI-TOF) (*see* **Note 5**).

3.3 Fluorescence In Vivo Hybridization (FIVH)

3.3.1 FIVH in Mice

Perform all procedures at room temperature and under sterile conditions unless otherwise specified. The FIVH can be performed 15 days post-infection (*see* **Note 6**). In each experiment, the following negative controls groups should be included: uninfected control group that received probe solution, to exclude false positive, and

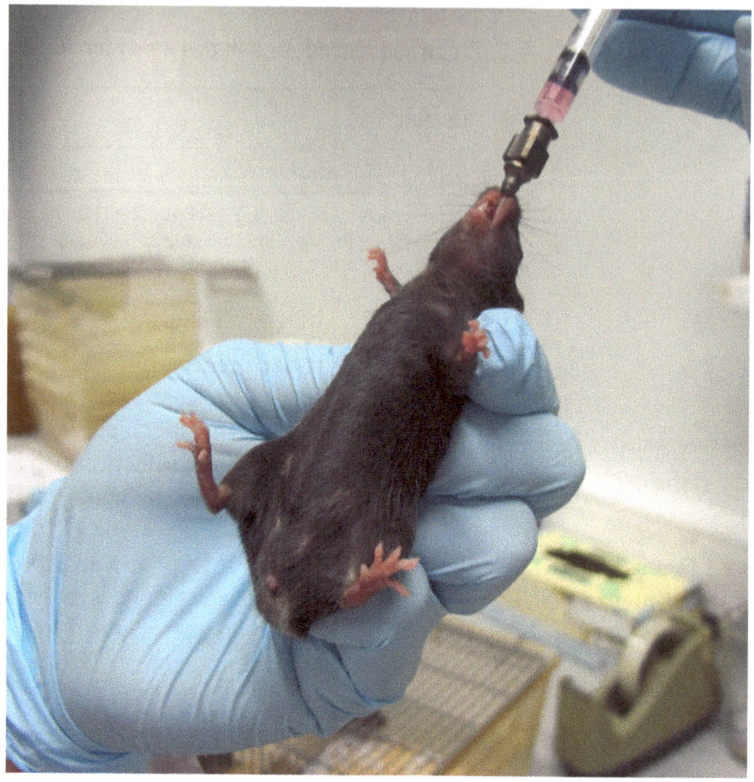

Fig. 2 Oral gavage performed in a C57BL/6 mouse. Probe diluted in the adjuvant buffer should be administered directly into the stomach by oral gavage. A bulb tipped gastric gavage needle or a flexible cannula or tube is attached to a syringe and used to deliver the compound into the stomach. During the whole procedure the animal is restrained, as it is shown

an infected control group that received the vehicle probe solution without probe, to evaluate specificity.

1. Dilute the Cy3-labeled probe in adjuvant buffer at 2 μM concentration (*see* **Notes 7** and **8**).
2. Administer the probe diluted in adjuvant buffer by oral gavage, using a 1 ml syringe coupled to a bulb tipped gastric gavage needle or a flexible polyethylene catheter (Fig. 2) (*see* **Note 9**).
3. Sacrifice animals after 30 min by CO_2 overdose (*see* **Note 9**).
4. Necropsy of each mouse. Remove the stomach in aseptic conditions and open it along the greater curvature.
5. After opening the stomach, recover the mucus with a coverslip, without pressuring the tissue.
6. Wash the glandular stomach in phosphate-buffered saline (PBS, pH 7.4) and divide it into tissue fragments representing cardia, body, and antrum regions.

7. Use half of the stomach for *H. pylori* culture (colony-forming units or CFU) (*see* **Note 10**). This will allow assessing the colonization efficiency, i.e., the presence and concentration of *H. pylori* bacteria.
8. Divide the remaining half of the glandular stomach into two parts. Rinse one part in PBS with 0.01% (v/v) NaN_3 and immediately freeze it in liquid nitrogen in optimal cutting temperature compound (OCT) for histopathological examination and analysis of fluorescence.
9. Fix the other part of the stomach in 4% (v/v) paraformaldehyde (PAF), for 1 h, at room temperature. Afterward, wash the tissue three times in PBS and store it in PBS with 0.01% (v/v) NaN_3, at 4 °C until further processing and paraffin embedding.
10. Prepare tissue cryosections with 5–10 μm thickness using a cryostat. Keep the samples at 80 °C, until processing for microscopy analysis.
11. Prepare paraffin tissue sections with 3 μm using a microtome (*see* **Note 11**). Keep the samples at −20 °C, until the analysis by microscopy.

3.3.2 Analysis of the Samples by Epifluorescence Microscopy

The detection of FIVH signal may be performed ex vivo in mucus samples or in paraffin-embedded sections, and cryosections.

1. Evaluate mucus, cryosections, and paraffin-embedded sections in a confocal system equipped with 405, 488, and 561 nm diode lasers for excitation of blue, green, and red fluorochromes, respectively. Take images in all the filters to exclude artifacts in the images. The analysis of the antrum and the corpus should be performed with more detail, as they are the preferential sites of *H. pylori* infection.
2. Acquire images with equal exposure times, for fluorescence intensity comparisons.
3. Analyze the acquired images, using image analysis software (*see* **Note 12**).

4 Notes

1. For other microorganisms or in case of if a different hybridization temperature is used, it will be necessary to perform the design of new probes.
2. LNA and 2′OMe monomers and probes are commercially available (e.g., Exiqon and Ribotask, respectively).
3. The selection of the fluorochrome needed to be based on the lower background obtained in tissues. High background sig-

nals are usually obtained in the green (488 nm) and blue (358 nm) channels.

4. The coupling yields should be between 95–99% per step to not compromise the synthesis yield.
5. The purity of the probe should be >90% to not compromise the specificity of FIVH.
6. The time after infection necessary to perform FIVH is dependent on the mouse, and bacteria species, and on the procedure selected for the infection.
7. The reaction stringency can be adjusted by parameters such as salt buffers or denaturants components.
8. Regarding the stomach environment, the integrity, sensitivity, and specificity of the probe has to be previously analyzed using acid pH. Additionally, the cytotoxicity of the probe should be studied on a gastric cells line through the analysis of cell viability and apoptosis induction. The analysis of the genotoxicity of the probe and the adjuvant buffer should be also tested previously, by standard assays, such as VITOTOX assay®.
9. Animals should be handled by trained and experienced personnel for routine maintenance and for the experiments.
10. To determine the number of colony-forming units (CFU) from tissues, each tissue fragment was weighed and homogenized in 1 ml TSB, using a TissueRuptor. Dilute the homogenate (serial dilutions) and plate in duplicate onto *H. pylori* selective growth medium, Tryptic Soy Agar (TSA) plates, supplemented with 5% sheep blood, vancomycin (10 μg/ml), trimethroprim (5 μg/ml), amphotericin (5 μg/ml), and cefsulodin (10 μg/ml) (Sigma-Aldrich), and incubate at 37 °C, under microaerophilic conditions. After 5 days of incubation, identify (colony morphology) and count the CFU per gram of stomach.
11. To facilitate the visualization transverse sections should be performed.
12. The detection of bacteria in the gastric epithelium indicates an efficient diffusion of the probe through the mucus layer.

Acknowledgments

This work was financially supported by Project UID/EQU/00511/2013-LEPABE, by the FCT/MEC with national funds and when applicable cofunded by FEDER in the scope of the P2020 Partnership Agreement; by FEDER funds through Programa Operacional Factores de Competitividade—COMPETE, by the Programa Operacional do Norte (ON2) program, Project

NORTE-07-0124-FEDER-000022 to M.L., and by national funds through FCT (Fundação para a Ciência e a Tecnologia): DNA mimics Research Project PIC/IC/82815/2007, Ph.D. grant, SFRH/BD/72999/2010 to S.F., and Post-Doctoral fellowship SFRH/BPD/33420/2008 to M.L.

References

1. DeLong EF, Wickham GS, Pace NR (1989) Phylogenetic stains: ribosomal RNA-based probes for the identification of single cells. Science 243(4896):1360–1363
2. Guimaraes N, Azevedo NF, Figueiredo C, Keevil CW, Vieira MJ (2007) Development and application of a novel peptide nucleic acid probe for the specific detection of Helicobacter pylori in gastric biopsies. J Clin Microbiol 45(9):3089–3094
3. Wagner M, Horn M, Daims H (2003) Fluorescence in situ hybridisation for the identification and characterisation of prokaryotes. Curr Opin Microbiol 6(3):302–309
4. Amann R, Fuchs BM (2008) Single-cell identification in microbial communities by improved fluorescence in situ hybridization techniques. Nat Rev Microbiol 6(5):339–348. doi:10.1038/nrmicro1888
5. Amann RI, Ludwig W, Schleifer KH (1995) Phylogenetic identification and in situ detection of individual microbial cells without cultivation. Microbiol Rev 59(1):143–169
6. Amann RI, Krumholz L, Stahl DA (1990) Fluorescent-oligonucleotide probing of whole cells for determinative, phylogenetic, and environmental studies in microbiology. J Bacteriol 172(2):762–770
7. Makristathis A, Riss S, Hirschl AM (2014) A novel fluorescence in situ hybridization test for rapid pathogen identification in positive blood cultures. Clin Microbiol Infect 20(10):O760–O763. doi:10.1111/1469-0691.12561
8. Harris DM, Hata DJ (2013) Rapid identification of bacteria and Candida using PNA-FISH from blood and peritoneal fluid cultures: a retrospective clinical study. Ann Clin Microbiol Antimicrob 12:2. doi:10.1186/1476-0711-12-2. 1476-0711-12-2 [pii]
9. Quast C, Pruesse E, Yilmaz P, Gerken J, Schweer T, Yarza P, Peplies J, Glockner FO (2013) The SILVA ribosomal RNA gene database project: improved data processing and web-based tools. Nucleic Acids Res 41(Database issue):D590–D596. doi:10.1093/nar/gks1219
10. Cole JR, Wang Q, Fish JA, Chai B, McGarrell DM, Sun Y, Brown CT, Porras-Alfaro A, Kuske CR, Tiedje JM (2014) Ribosomal Database Project: data and tools for high throughput rRNA analysis. Nucleic Acids Res 42(Database issue):D633–D642. doi:10.1093/nar/gkt1244. gkt1244 [pii]
11. Zhang H (2003) Alignment of BLAST high-scoring segment pairs based on the longest increasing subsequence algorithm. Bioinformatics 19(11):1391–1396
12. Berlier JE, Rothe A, Buller G, Bradford J, Gray DR, Filanoski BJ, Telford WG, Yue S, Liu J, Cheung CY, Chang W, Hirsch JD, Beechem JM, Haugland RP, Haugland RP (2003) Quantitative comparison of long-wavelength Alexa Fluor dyes to Cy dyes: fluorescence of the dyes and their bioconjugates. J Histochem Cytochem 51(12):1699–1712
13. Hayashi-Takanaka Y, Stasevich TJ, Kurumizaka H, Nozaki N, Kimura H (2014) Evaluation of chemical fluorescent dyes as a protein conjugation partner for live cell imaging. PLoS One 9(9):e106271. doi:10.1371/journal.pone.0106271
14. Hugenholtz P, Tyson GW, Blackall LL (2002) Design and evaluation of 16S rRNA-targeted oligonucleotide probes for fluorescence in situ hybridization. Methods Mol Biol 179:29–42
15. Cerqueira L, Azevedo NF, Almeida C, Jardim T, Keevil CW, Vieira MJ (2008) DNA mimics for the rapid identification of microorganisms by fluorescence in situ hybridization (FISH). Int J Mol Sci 9(10):1944–1960. doi:10.3390/ijms9101944
16. Kubota K, Ohashi A, Imachi H, Harada H (2006) Improved in situ hybridization efficiency with locked-nucleic-acid-incorporated DNA probes. Appl Environ Microbiol 72(8):5311–5317. doi:10.1128/aem.03039-05
17. Robertson KL, Vora GJ (2012) Locked nucleic acid and flow cytometry-fluorescence in situ hybridization for the detection of bacterial small noncoding RNAs. Appl Environ Microbiol 78(1):14–20. doi:10.1128/aem.06399-11
18. Fontenete S, Guimaraes N, Leite M, Figueiredo C, Wengel J, Filipe Azevedo N

(2013) Hybridization-based detection of Helicobacter pylori at human body temperature using advanced locked nucleic acid (LNA) probes. PLoS One 8(11):e81230. doi:10.1371/journal.pone.0081230. PONE-D-13-15277 [pii]

19. Fontenete S, Leite M, Guimaraes N, Madureira P, Ferreira RM, Figueiredo C, Wengel J, Azevedo NF (2015) Towards Fluorescence In Vivo Hybridization (FIVH) detection of H. pylori in gastric mucosa using advanced LNA probes. PLoS One 10(4):e0125494. doi:10.1371/journal.pone.0125494. PONE-D-14-54901 [pii]
20. Fontenete SLM, Cappoen D, Santos R, Ginneken CV, Figueiredo C, Wengel J, Cos P, Azevedo NF (2016) Fluorescence in vivo hybridization (FIVH) for detection of Helicobacter pylori infection in a C57BL/6 mouse model. PLoS One 11(2):e0148353
21. Marshall BJ (1988) The Campylobacter pylori story. Scand J Gastroenterol Suppl 146:58–66
22. Lee A (1994) The microbiology and epidemiology of Helicobacter pylori infection. Scand J Gastroenterol Suppl 201:2–6
23. Fontenete S, Leite M, Cappoen D, Santos R, Ginneken CV, Figueiredo C, Wengel J, Cos P, Azevedo NF (2016) Fluorescence In Vivo Hybridization (FIVH) for detection of Helicobacter pylori infection in a C57BL/6 mouse model. PLoS One 11(2):e0148353. doi:10.1371/journal.pone.0148353. PONE-D-15-52300 [pii]
24. Kumar R, Singh SK, Koshkin AA, Rajwanshi VK, Meldgaard M, Wengel J (1998) The first analogues of LNA (locked nucleic acids): phosphorothioate-LNA and 2′-thio-LNA. Bioorg Med Chem Lett 8(16):2219–2222. doi:S0960894X98003667 [pii]
25. You Y, Moreira BG, Behlke MA, Owczarzy R (2006) Design of LNA probes that improve mismatch discrimination. Nucleic Acids Res 34(8):e60. doi:10.1093/nar/gkl175. 34/8/e60 [pii]
26. Kierzek E, Ciesielska A, Pasternak K, Mathews DH, Turner DH, Kierzek R (2005) The influence of locked nucleic acid residues on the thermodynamic properties of 2′-O-methyl RNA/RNA heteroduplexes. Nucleic Acids Res 33(16):5082–5093. doi:10.1093/nar/gki789. 33/16/5082 [pii]
27. Yan Y, Yan J, Piao X, Zhang T, Guan Y (2012) Effect of LNA- and OMeN-modified oligonucleotide probes on the stability and discrimination of mismatched base pairs of duplexes. J Biosci 37(2):233–241
28. Maciaszek A, Krakowiak A, Janicka M, Tomaszewska-Antczak A, Sobczak M, Mikolajczyk B, Guga P (2015) LNA units present in the (2′-OMe)-RNA strand stabilize parallel duplexes (2′-OMe)-RNA/[All-RP-PS]-DNA and parallel triplexes (2′-OMe)-RNA/[All-RP-PS]-DNA/RNA. An improved tool for the inhibition of reverse transcription. Org Biomol Chem 13(8):2375–2384. doi:10.1039/c4ob02364a
29. Fontenete S, Guimaraes N, Wengel J, Azevedo NF (2015) Prediction of melting temperatures in fluorescence in situ hybridization (FISH) procedures using thermodynamic models. Crit Rev Biotechnol 36(3):566–577. doi:10.3109/07388551.2014.993589
30. Adams AM, Harding PL, Iversen PL, Coleman C, Fletcher S, Wilton SD (2007) Antisense oligonucleotide induced exon skipping and the dystrophin gene transcript: cocktails and chemistries. BMC Mol Biol 8:57. doi:10.1186/1471-2199-8-57

Chapter 9

Rapid Antibiotic Susceptibility Testing for Urinary Tract Infections

Anja Mezger, Mats Nilsson, and Dan I. Andersson

Abstract

Antibiotic susceptibility testing is important to guide clinicians in their choice of antibiotic used for treatment of bacterial infections. Current methods are time-consuming and more rapid alternatives are needed. Here, we describe a novel rapid method for antibiotic susceptibility testing which combines phenotypic and genotypic measurements. The use of padlock probes and rolling circle amplification allows for fast and precise determination of antibiotic susceptibilities as well as species identification.

Key words Antibiotic susceptibility testing, Urinary tract infections, Padlock probes, Rolling circle amplification

1 Introduction

Rapid antibiotic susceptibility testing (AST) is essential for correct treatment in a timely fashion. Often physicians will prescribe a broad-spectrum antibiotic before the antibiotic susceptibility profile is known. Together with antibiotic overuse, this leads to an increase in antibiotic resistance and as a result it has generated one of the biggest medical challenges of our time [1]. Thus, there is an urgent need for the development of rapid AST methods. Currently, the most common method for AST is either agar or broth dilution where the minimum inhibitory concentration is measured or a disk diffusion assay where zone diameters are measured [2, 3]. Although technically simple, these methods require at least one overnight incubation for fast growing organisms, and up to several weeks for slow-growing bacteria such as *Mycobacterium tuberculosis.* Commercially available instruments for automated AST can significantly reduce turnaround time but with a lower identification rate if the inoculum is directly taken from urine samples compared to from pathogens isolated from chromogenic media [4].

Several new molecular methods have been developed for rapid AST such as various PCR assays and lately also MALDI-TOF has

Kimberly A. Bishop-Lilly (ed.), *Diagnostic Bacteriology: Methods and Protocols*, Methods in Molecular Biology, vol. 1616, DOI 10.1007/978-1-4939-7037-7_9, © Springer Science+Business Media LLC 2017

been used for the detection of antibiotic resistances [5–9]. PCR assays are fast and sensitive, but the resistance genes must be known and emerging resistance mechanisms will not be detected. At the same time, genotypic resistance does not necessarily correspond to phenotypic resistance [10] and thus, sole reliance on genotypes can lead to erroneous treatments. MALDI-TOF is currently still under development for the detection of antibiotic resistance and so far can only be used for the detection of a few classes of antibiotics and requires a pure culture as starting material, delaying test results [8].

In Mezger et al. we combined phenotypic and genotypic methods resulting in a rapid and sensitive novel assay for reliable AST [11]. Urinary samples were diluted in Lysogeny broth (LB) and cultured in the presence and absence of antibiotics for 2 h. Antibiotic resistance was defined by the relative bacterial growth in antibiotic-containing medium compared to non-antibiotic-containing medium. The chosen threshold for resistance might vary with the antibiotic substance tested and needs to be determined empirically. We quantitatively measured the DNA amount (which is directly proportional to growth) in each sample using species-specific padlock probes and rolling circle amplification (RCA) to calculate growth. Padlock probes are short oligonucleotides with target-complementary end sequences that are connected by a backbone sequence containing restriction sites and binding sites for detection of oligonucleotides [12]. After hybridization to their target sequence padlock probes are end-joined by enzymatic ligation and amplified by RCA [12–14]. To increase the amplification factor, rolling circle products are monomerized by restriction enzyme digestion, recircularized by ligation, and further amplified by RCA [15]. Rolling circle products spontaneously collapse into micrometer-sized coils and can be optically detected after hybridization of short fluorescently labeled oligonucleotides [14, 16]. We used a commercially available fluorescence detector to quantify the rolling circle products (Q-Linea AB, Sweden). The aqueous solution of rolling circle products is pumped through a microfluidic channel, imaged and counted.

2 Materials

2.1 Sample Pretreatment

1. Lysogeny broth. Store at 4 °C.
2. Antibiotics to be tested, e.g., ciprofloxacin (0.5 μg/ml final concentration) and trimethoprim (4 μg/ml final concentration).
3. Lysis solution: 5 M NaOH (*see* **Note 1**). Store at room temperature (RT).

4. Neutralization solution: 0.75 M Tris–HCl (pH 8), 2.5 M HCl (*see* **Note 2**). Store at RT.

2.2 Padlock Probe Phosphorylation

1. T4 Polynucleotide kinase (PNK).
2. 10× reaction buffer A for T4 Polynucleotide Kinase (50 mM Tris–HCl [pH 7.6], 10 mM $MgCl_2$, 5 mM dithiothreitol [DTT], 0.1 MM spermidine) (*see* **Note 3**).
3. ATP.

2.3 Padlock Probe Ligation, Amplification, and Detection

1. Hybridization buffer: 1.7 nM target-complementary capture oligonucleotide (Table 1), 10 mM Tris–HCl (pH 7.5), 5 mM EDTA, 0.1% Tween 20, 1 M NaCl (*see* **Note 4**). Store at RT, oligonucleotides are stored at −20 °C and added immediately prior to use.
2. Dynabeads MyOne streptavidin T1 (Life Technologies, Waltham, MA, USA). Store at 4 °C.
3. Washing buffer 1: 1 mM Tris–HCl (pH 7.5), 5 mM EDTA, 0.1% Tween 20, 0.1 mM NaCl. Store at RT.
4. Hybridization and ligation solution: 100 nM of each padlock probe (Table 1), 0.2 mg/ml BSA (New England Biolabs, Ipswich, MA, USA), 1× Ampligase reaction buffer (20 mM Tris–HCl [pH 8.3], 25 mM KCl, 10 mM $MgCl_2$, 0.5 mM NAD, 0.01% Triton X-100; Epicentre, Madison, WI, USA), 5 U Ampligase (Epicentre). Prepare freshly.
5. Washing buffer 2: 20 mM Tris–HCl (pH 8.3), 25 mM KCl, 10 mM $MgCl_2$, and 0.01% Triton X-100. Store at RT.
6. 1× phi29 DNA polymerase reaction buffer (33 mM Tris-acetate [pH 7.9], 10 mM Mg-acetate, 66 mM K-acetate, 0.1% Tween 20, 1 mM DTT; Thermo Scientific).
7. RCA solution 1: 25 nM primer, 0.2 mg/ml BSA, 125 μM dinucleotides (dNTPs), 1× phi29 DNA polymerase reaction buffer, 100 mU/μl phi29 DNA polymerase (Thermo Scientific). Prepare freshly.
8. Restriction digestion solution: 0.2 mg/ml BSA, 1× phi29 DNA polymerase reaction buffer, 600 mU/μl AluI, 600 nM restriction oligonucleotides (Table 1). Prepare freshly.
9. RCA solution 2: 0.2 mg/ml BSA, 1.36 mM ATP, 28 mU/μl T4 DNA ligase, 1× phi29 DNA polymerase reaction buffer, 10 nM of each detection oligonucleotide (Table 1), 120 mU/μl phi29 DNA polymerase. Prepare freshly.
10. Labeling buffer: 25 nM of each detection oligonucleotide, 10 mM EDTA, 10 mM Tris–HCl [pH 7.5], 0.25% Tween 20, 0.5 M NaCl.

Table 1
Oligonucleotide sequences used in the assay

Name	Sequence
Escherichia coli padlock probe	5′-TTAATACCTTTGCTCATTGACAGAGTGTATGCAGCTCCTCAGTA TAGTCGATAGTCACGGCTACTTTTGGAAGGGAGTAAAG-3′
Proteus mirabilis padlock probe	5′-GACCTTGCACTATCGGATGAGAGTGTATGCAGCTCCTCAGTATA GTCGATAGTCACGGCTACTTTTGGGGCTCTTCG-3′
Pseudomonas aeruginosa padlock probe	5′-AGGGAGAAAGTGAGAGTGTATGCAGCTCCTCAGTATAGTCGAT AGTCACGGCTACTTTTCCGCATACGTCCTG-3′
E. coli capture oligonucleotide	5′-CTCTCTCTCTCTCTCTCTCTCTCTCTGAAGAAGCACCGGCTAACT CCGTGCCAGCAGCCGCGGTAA-3′
P. mirabilis capture oligonucleotide	5′-CTCTCTCTCTCTCTCTCTCTCTCTCTGGATTAGCTAGTAGGTGGG GTAAAGGCTCACCTAGGCGAC-3′
P. aeruginosa capture oligonucleotide	5′-CTCTCTCTCTCTCTCTCTCTCTCTCTCGGACCTCACGCTATCAGA TGAGCCTAGGTCGGATTAGCTA-3′
Primer	5′-TACTGAGGAGCTGCATACAC-3′
Restriction oligonucleotide	5′-GTGTATGCAGCTCCTCAGTA-3′
Detection oligonucleotide 1	Cy3-5′-AG+TA+GC+CG+TGTTC**UUUU**-3′
Detection oligonucleotide 2	Cy3-5′-AC+TA+TC+GA+CTTTC**UUUU**-3′[a]

[a] + before a base indicates LNA residues; bases in bold are 2′OMe-RNA

3 Methods

3.1 Sample Pretreatment

1. Split each sample into the number of tested antibiotics plus one for the non-antibiotic containing control (*see* **Note 5**). Add 10 μl of urine to 90 μl of LB either supplemented with antibiotics or without antibiotics into a 96-well plate. Culture the bacteria for 2 h at 37 °C (*see* **Note 6**).
2. Add 10 μl of 5 M NaOH to each sample to lyse the bacteria and incubate at 95 °C for 10 min.
3. Adjust the pH by adding 20 μl of the neutralization solution.

3.2 Padlock Probe Phosphorylation

1. Add padlock probes to a final concentration of 1 μM to 1× PNK buffer A, 1 mM ATP, and 10 U of T4 PNK. Incubate the solution at 37 °C for 30 min and inactivate the enzyme at 65 °C for 20 min. Phosphorylate each padlock probe separately.

3.3 Padlock Probe Ligation, Amplification, and Detection

1. Denature the DNA by incubating samples at 98 °C for 5 min, followed by snap-cooling on ice.
2. Add 47 μl of hybridization solution to each sample and incubate at 55 °C for 10 min. The biotinylated target-complementary capture oligonucleotides (Table 1) will bind to their target sequence and can be captured onto magnetic beads in the following step.
3. Add 10 μl of Dynabeads MyOne streptavidin T1 beads (*see* **Note 7**) to each sample and incubate at RT for 10 min.
4. Wash the samples once with 100 μl washing buffer 2 using a 96-well plate magnet separator.
5. Keep the samples on the plate magnet separator and replace the liquid with 20 μl of the hybridization and ligation solution. To allow efficient hybridization and ligation, incubate the samples at 55 °C for 5 min.
6. Wash the samples once with 100 μl washing buffer 1.
7. Retain the beads using the plate magnet separator and replace the liquid with 20 μl of RCA solution 1. Incubate the samples at 37 °C for 20 min for amplification of ligated padlock probes and inactivate the polymerase by incubating the samples at 65 °C for 1 min.
8. Enzymatically digest the rolling circle products by adding 5 μl of restriction digestion solution and incubate the samples at 37 °C for 1 min, followed by 65 °C for 1 min to inactivate the restriction enzyme.

9. Add 25 μl of RCA solution 2 to further amplify the target sequences. Incubate samples at 37 °C for 20 min, followed by 65 °C for 1 min.
10. Dilute the samples 1:100 in labeling buffer and count the number of rolling circle products by either using the Aquila400 or using a conventional epifluorescence microscope. If using an epifluorescence microscope a fraction of the samples can be spotted onto a glass slide.

4 Notes

1. Sodium hydroxide may be corrosive to metals and appropriate personal protective equipment should be worn. Sodium hydroxide can cause skin burns and eye damage.
2. Hydrochloric acid is highly corrosive and appropriate personal protective equipment should be worn. Hydrochloric acid can cause skin burns, eye damage, and respiratory irritation.
3. DTT may be precipitated in the reaction buffer A for T4 Polynucleotide Kinase after thawing it. Warming it to 45 °C prior to use will dissolve any precipitates.
4. Prepare all solutions using ultrapure water free of DNases and keep them cooled during experimental preparations. All solutions and enzymes should be stored at −20 °C unless otherwise indicated.
5. Carry out the sample pretreatment in a BSL-2 lab and wear appropriate personal protective equipment to avoid any accidental infection. All other steps can be carried out in a BSL-1 lab.
6. We find that it is most convenient to carry out all incubations in a PCR instrument.
7. Wash the magnetic beads three times with 100 μl washing buffer 1 prior to use.

References

1. Costelloe C, Metcalfe C, Lovering A, Mant D, Hay AD (2010) Effect of antibiotic prescribing in primary care on antimicrobial resistance in individual patients: systematic review and meta-analysis. BMJ 340:c2096. doi:10.1136/bmj.c2096. bmj.c2096 [pii]
2. Wiegand I, Hilpert K, Hancock RE (2008) Agar and broth dilution methods to determine the minimal inhibitory concentration (MIC) of antimicrobial substances. Nat Protoc 3(2):163–175. doi:10.1038/nprot.2007.521. nprot.2007.521 [pii]
3. CLSI (2015) Performance standards for antimicrobial susceptibility testing; twenty-fifth informational supplement (M100-S25). Clinical and Laboratory Standards Institute, Wayne, PA
4. Munoz-Davila MJ, Roig M, Yague G, Blazquez A, Salvador C, Segovia M (2013) Comparative evaluation of Vitek 2 identification and susceptibility testing of urinary tract pathogens directly and isolated from chromogenic media. Eur J Clin Microbiol Infect Dis 32(6):773–780. doi:10.1007/s10096-012-1806-4

5. Cuny C, Witte W (2005) PCR for the identification of methicillin-resistant *Staphylococcus aureus* (MRSA) strains using a single primer pair specific for SCCmec elements and the neighbouring chromosome-borne orfX. Clin Microbiol Infect 11(10):834–837 . doi:10.1111/j.1469-0691.2005.01236.xCLM1236 [pii]
6. Brolund A, Wisell KT, Edquist PJ, Elfstrom L, Walder M, Giske CG (2010) Development of a real-time SYBRGreen PCR assay for rapid detection of acquired AmpC in enterobacteriaceae. J Microbiol Methods 82(3):229–233. doi:10.1016/j.mimet.2010.06.006. S0167-7012(10)00203-4 [pii]
7. Oviano M, Fernandez B, Fernandez A, Barba MJ, Mourino C, Bou G (2014) Rapid detection of enterobacteriaceae producing extended spectrum beta-lactamases directly from positive blood cultures by matrix-assisted laser desorption ionization-time of flight mass spectrometry. Clin Microbiol Infect 20(11):1146–1157. doi:10.1111/1469-0691.12729
8. Hrabak J, Chudackova E, Walkova R (2013) Matrix-assisted laser desorption ionization-time of flight (maldi-tof) mass spectrometry for detection of antibiotic resistance mechanisms: from research to routine diagnosis. Clin Microbiol Rev 26(1):103–114. doi:10.1128/CMR.00058-12. 26/1/103 [pii]
9. Kempf M, Bakour S, Flaudrops C, Berrazeg M, Brunel JM, Drissi M, Mesli E, Touati A, Rolain JM (2012) Rapid detection of carbapenem resistance in Acinetobacter Baumannii using matrix-assisted laser desorption ionization-time of flight mass spectrometry. PLoS One 7(2):e31676. doi:10.1371/journal.pone.0031676. PONE-D-11-22833 [pii]
10. Vilhelmsson SE, Tomasz A, Kristinsson KG (2000) Molecular evolution in a multidrug-resistant lineage of *Streptococcus pneumoniae*: emergence of strains belonging to the serotype 6B Icelandic clone that lost antibiotic resistance traits. J Clin Microbiol 38(4): 1375–1381
11. Mezger A, Gullberg E, Goransson J, Zorzet A, Herthnek D, Tano E, Nilsson M, Andersson DI (2015) A general method for rapid determination of antibiotic susceptibility and species in bacterial infections. J Clin Microbiol 53(2):425–432. doi:10.1128/JCM.02434-14. JCM.02434-14 [pii]
12. Nilsson M, Malmgren H, Samiotaki M, Kwiatkowski M, Chowdhary BP, Landegren U (1994) Padlock probes: circularizing oligonucleotides for localized DNA detection. Science 265(5181):2085–2088
13. Baner J, Nilsson M, Mendel-Hartvig M, Landegren U (1998) Signal amplification of padlock probes by rolling circle replication. Nucleic Acids Res 26(22):5073–5078. doi:gkb813 [pii]
14. Lizardi PM, Huang X, Zhu Z, Bray-Ward P, Thomas DC, Ward DC (1998) Mutation detection and single-molecule counting using isothermal rolling-circle amplification. Nat Genet 19(3):225–232. doi:10.1038/898
15. Dahl F, Baner J, Gullberg M, Mendel-Hartvig M, Landegren U, Nilsson M (2004) Circle-to-circle amplification for precise and sensitive DNA analysis. Proc Natl Acad Sci U S A 101(13):4548–4553. doi:10.1073/pnas.0400834101. 0400834101 [pii]
16. Jarvius J, Melin J, Goransson J, Stenberg J, Fredriksson S, Gonzalez-Rey C, Bertilsson S, Nilsson M (2006) Digital quantification using amplified single-molecule detection. Nat Methods 3(9):725–727. doi:10.1038/Nmeth916

Chapter 10

Detection and Differentiation of Lyme Spirochetes and Other Tick-Borne Pathogens from Blood Using Real-Time PCR with Molecular Beacons

Samantha Schlachter, Kamfai Chan, Salvatore A.E. Marras, and Nikhat Parveen

Abstract

Real-time PCR assays have recently been implemented in diagnostics for many bacterial pathogens, allowing rapid and accurate detection, which ultimately results in improved clinical intervention. Here, we describe a sensitive method of detection for three common tick-borne pathogens *Borrelia burgdorferi*, *Anaplasma phagocytophilum*, and *Babesia microti* since coinfections with these pathogens have started occurring with increasing frequency over the last several years in both North America and Europe. A shared geographic region, the same tick vectors, and similar transmission cycle all favor simultaneous transmission of these three tick-borne pathogens. Furthermore, early symptoms of the diseases are often similar and somewhat nonspecific leading to poor clinical identification. The multiplex real-time PCR assay we describe here utilizes gene-specific primers, molecular beacon probes tagged with different fluorophores, and optimized PCR conditions to detect even small amounts of specific pathogen DNA without interference. Application of this detection method will offer better diagnostics for acute and persistent infection compared to the two-tier serological tests that are currently approved in North America and Europe, which do not necessarily detect active infection.

Key words Mutliplex assay, Molecular beacons, Real-time PCR, Tick-borne pathogens, *Borrelia burgdorferi*, *Anaplasma phagocytophilium*, *Babesia microti*, Lyme disease, Babesiosis, Anaplasmosis

1 Introduction

As the frequency of tick-borne disease continues to rise worldwide, researchers and clinicians alike seek to find more sensitive and specific detection methods for diagnosis. The current diagnostic paradigm for Lyme disease and other tick-borne pathogens depends either on labor-intensive microscopic examination of blood smears or on serological tests, which are not useful either in the early stages of infection, prior to the establishment of an adaptive immune response, or after the pathogen is cleared due to antibody persistence [1–3]. Further complicating diagnosis of a specific tick-borne

Kimberly A. Bishop-Lilly (ed.), *Diagnostic Bacteriology: Methods and Protocols*, Methods in Molecular Biology, vol. 1616, DOI 10.1007/978-1-4939-7037-7_10,

disease is the possibility of coinfection with multiple pathogens. The same vector, overlapping geographical regions, and transmission cycles of *Borrelia burgdorferi*, *Anaplasma phagocytophilum*, and *Babesia microti*, which cause Lyme disease, anaplasmosis, and babesiosis, respectively often result in coinfection [4, 5]. The inoculation of multiple pathogens can alter the host immune response and change pathogenesis [6]. Coinfections are frequently associated with more severe disease and persistent complications [7]. Therefore, it will be highly beneficial to develop diagnostic techniques to detect the three pathogens simultaneously. Here, we describe a sensitive multiplex assay for the detection of *Borrelia species*, *A. phagocytophilum*, and *B. microti* and potentially other *Babesia* spp.

Real-time PCR (rtPCR)-based diagnostic assays have been implemented clinically for the diagnosis of many different pathogen-associated diseases (reviewed in [8]). For instance, the rapid detection of nucleic acids from *Staphylococcus*, *Streptococcus*, *Mycobacterium* and zoonotic pathogens such as *Bartonella* and *Yersinia*, by rtPCR has allowed rapid and enhanced clinical intervention [9–14]. Selection of proper target amplicons and conditions along with the use of efficient detection probes are critical for the development of sensitive and specific PCR-based diagnostic assays. The fundamental advantage of rtPCR is the ability to detect not only the presence of a specific pathogen with gene-specific primers and molecular beacon probes, but also to distinguish multiple pathogens in real time [15]. Our specific assay for *B. burgdorferi*, *A. phagocytophilum*, and *B. microti* utilizes primers to amplify regions of the *recA*, APH1387 and TPK genes, respectively [16]. The amplicon from each pathogen is detected by a unique molecular beacon probe conjugated with a distinguishable colored fluorophore, thus allowing for detection through a multiplex assay approach. Molecular beacons are oligonucleotide probes that contain a fluorophore and quencher separated by a single-stranded loop region that is specific for the target amplicon sequence, additionally the stem structure contains random regions of complementary nucleotides that keep the fluorophore and quencher in direct association [17, 18]. This direct contact is very efficient and prevents background fluorescence. It is only upon the formation of a probe-target hybrid, and the resulting physical disruption of the interaction between the quencher and fluorophore that fluorescence is emitted [15, 17, 19]. A summary of specific molecular beacon design and various applications can be found at www.molecular-beacons.org. Previous efforts at using rtPCR for the detection of *B. burgdorferi* and other tick-borne pathogens have been limited in their specificity, however, by implementing molecular beacons, which have been shown to distinguish single nucleotide polymorphisms; we are able to detect even very low levels of infection [16, 17, 20, 21].

The molecular beacons employed in our multiplex rtPCR assay can also distinguish between three species of *B. burgdorferi* sensu lato all of which have been shown to cause Lyme disease in humans. *B. burgdorferi* sensu stricto is primarily responsible for causing Lyme disease in North America, while *B. afzelli* and *B. garinii* are also prevalent in Europe [22]. The *recA* gene sequence of *Borrelia* differs slightly among these three species, which alters the melting temperature (Tm) of the molecular beacon probe-target hybrid. The Tm for *B. burgdorferi recA* amplicon, for which the molecular beacon probe was designed, is 71 °C, while sequence variations between *B. afzelii* and *B. garinii* result in weaker probe-target interactions and therefore, the Tm decrease to 63 °C and 57 °C, respectively (Table 1, Fig. 1).

The rtPCR assay we describe here has clinical applications for the detection of three *Borrelia* species DNA in patient samples for improved Lyme diagnostics. A duplex assay is also useful to study Lyme disease pathogenesis in the murine model of infection by facilitating detection and quantification of *B. burgdorferi* DNA relative to mouse DNA copy number in various tissues. This allows for researchers to determine colonization defects of specific mutants generated in a spirochete strain. Furthermore, our multiplex approach will allow for rapid screening of both *Babesia* and *Anaplasma* species to prevent blood transfusion-associated infections, which pose a serious and even fatal risk to patients [23]. Therefore, protocols for DNA isolation and multiplex rtPCR analysis of both patient blood samples and infected mouse tissue are outlined. We provide additional evidence supporting the efficiency and sensitivity of rtPCR for the detection of tick-borne pathogens [16]. Lastly, we demonstrate for the first time that the storage of patient blood samples at four degrees Celsius for a week does not impair the sensitivity of the assay, thus allowing for blood collection and analysis to be carried out in a convenient manner.

2 Materials

2.1 DNA Isolation

1. QIAamp DNA Blood Midi Kit.
2. Qiagen DNeasy Blood and Tissue Kit.
3. 2 mL and 15 mL PLG-containing tubes (Qiagen MaXtract tubes).

Table 1
RecA3 for Borrelia species differentiation

Target/Probe	Sequence
Bb-RecA3	ttat GCGCCCCCTAGGATATCCGCCA atgc
Ba-RecA3	ttat GCGCCCCCTAGGATATCCACCA atgc
Bg-RecA3	ttat TCGCCCCCTAGGATATCCACCA atgc
RecA3 probe	5′ CTG **GCG GAT ATC CTA GGG GG CGC** CAG 3′

Bold letters indicate the probe sequence complementary to Bb-RecA3 target sequence

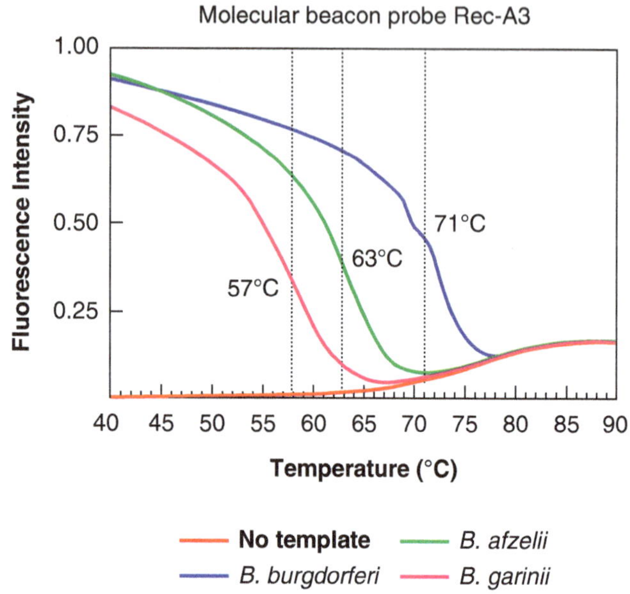

Fig. 1 Melt curve analysis following rtPCR with molecular beacons can differentiate *Borrelia* species. For *Borrelia* species identification rtPCR was immediately followed by melt curve analysis to identify the probe-target Tm value. Samples were incubated at 25 °C for 2 min to allow probe-target annealing. Subsequently, a melt curve was generated by increasing temperature from 25 °C to 95 °C in a 1 °C stepwise manner, with each step lasting 2 min, while monitoring the fluorescence. The Tm values for *B. burgdorferi*, *B. afzelii*, and *B. garinii* are 71 °C, 63 °C, and 57 °C, respectively, at which target and probe separate

4. Mouse Tissue Lysis Buffer (50 mM Tris–HCl, 100 mM NaCl, 100 mM EDTA): can be prepared ahead of time and stored at room temperature.
5. Proteinase K stock solutions: 50 mg/mL (for patient sample DNA isolation) or 20 mg/mL (for mouse tissue DNA isolation).
6. Solution 1: Phenol: chloroform: isoamyl alcohol mixture (25:24:1 ratio); the volume of solutions needed for phenol: chloroform extractions will vary depending on the number of samples from which DNA will be isolated.
7. Solution 2: Chloroform: isoamyl alcohol (24:1 ratio). Both solutions containing chloroform should be prepared fresh on the day of use.

2.2 Quantitative Real-Time PCR

1. Optical tubes with flat caps (0.2 mL) or semi-skirted 96-well PCR plates.
2. Thermal seal silicone adhesive film (transparent, non-fluorescent) if 96-well plates are used.

3. Duplex reaction mixture: 10× AmpliTaq Gold buffer, BSA (5 mg/mL), $MgCl_2$ (25 mM), dNTPs (25 mM), RecF primer (5 μM), RecR primer (5 μM), RecA3 probe (100 ng/μL), NidoF primer (5 μM), NidoR primer (5 μM), Nidogen probe (100 ng/μL), AmpliTaq Gold (5 U/μL).
4. Quadruplex reaction mixture: 10× AmpliTaq Gold buffer, BSA (5 mg/mL), $MgCl_2$ (25 mM), dNTPs (25 mM), RecF primer (50 μM), RecR primer (50 μM), RecA3 probe (100 ng/μL), 5BmTPK primer (50 μM), 3BmTPK primer (50 μM), BmTPK probe (100 ng/μL), 5Aphagocyt primer (50 μM), 3Aphagocyt primer (50 μM), Aph1387 probe (100 ng/μL), 5ACTA1 primer (5 μM), 3ACTA1 primer (5 μM), ACTA1 probe (100 ng/μL), AmpliTaq Gold (5 U/μL).

3 Methods

3.1 DNA Isolation

3.1.1 DNA Isolation from Patient Samples

1. Centrifuge whole blood samples at 10,000 × *g* for 5 min to separate the plasma from the cell pellet. Removed plasma can be stored at −80 °C.
2. Add one third of the volume of 50 mg/mL proteinase K. Add 4 μL of RNase A per every 1 mL of cell pellet.
3. Incubate all samples in a 55–60 °C water bath for 1 h. Vortex each tube every 5 min until cell lysis is complete and uniform suspension is obtained.
4. After lysis, use the QIAamp DNA Blood Midi Kit for patient samples to obtain the highest yield of best quality DNA. Add Buffer AL in a 1:1 ratio with lysate and vortex thoroughly. Then, incubate in a 70 °C water bath for 10 min.
5. Add equal volume of 100% ethanol (1:1 ratio with suspension in the tube). Vortex thoroughly.
6. Due to the large volume of blood that is typically collected, it is necessary to utilize the QIAamp DNA Blood Midi Kit. Transfer the lysate to the spin column provided with the kit. Briefly centrifuge at 6000 × *g* and discard the flow through.
7. Elute the DNA from the column by adding 400 μL of TE elution buffer, wait 5 min and centrifuge (*see* **Note 1**).
8. Add the eluent to the top of the column, wait 5 min and centrifuge again to recover concentrated DNA.
9. Add 400 μL of phenol to the eluent and mix by inversion, add 400 μL of chloroform and mix by inversion. Transfer the mixture to a 2 mL PLG-containing tube and centrifuge at 3500 × *g* for 10 min.
10. Collect the top aqueous layer in a clean microfuge tube and add 240 μL isopropanol to precipitate DNA, mix and let it

stand for 5 min. Transfer the mixture to a clean DNeasy kit spin column centrifuge at a maximum speed (21,000 × *g*) for 1 min. Discard flow through.

11. Add 500 μL AW1 and centrifuge at maximum speed for 1 min. Discard flow through.
12. Add 500 μL AW2 and centrifuge at maximum speed for 1 min. Discard flow through.
13. Spin the column at maximum speed for 3 min to dry the membrane. Discard flow through and collection tube.
14. Place the column into a clean microfuge tube and add 50–100 μL of nuclease-free water, wait 5 min and centrifuge (*see* **Note 1**).
15. Add the eluent to the top of the column, wait 5 min and centrifuge again to recover concentrated DNA.
16. Check the final DNA concentration by spectrophotometry (*see* **Note 2** for troubleshooting).

3.1.2 DNA Isolation from Infected Mouse Tissue

1. Centrifuge tissues at 2000 × *g* for 5 min at a temperature of 4 °C (*see* **Note 3**).
2. Add Tissue Lysis Buffer (50 mM Tris–HCl, 100 mM NaCl, 100 mM EDTA) enough to cover the entire tissue. Typically, 700 μL is sufficient for all mice tissues except for the animal joint, which requires double the volume, approximately 1.4 mL.
3. Add 1 mg/mL Qiagen Proteinase K. The volume of proteinase K should be determined based on the volume of lysis buffer added such that 35 μL of 20 mg/mL stock is added per 700 μL lysis buffer.
4. Vortex tissues for approximately 2 min and centrifuge at 2000 × g for 5 min at 4 °C.
5. Incubate samples in a 60 °C water bath overnight.
6. Vortex each tube for approximately 2 min next day to mix and return tubes to the water bath. Vortex hourly to homogenize the tissues (*see* **Note 4**).
7. Centrifuge at 2000 × *g* for 5 min at 4 °C. Incubate tubes in the water bath overnight.
8. Centrifuge tubes at 2000 × *g* for 5 min at 4 °C to remove undigested material.
9. Transfer the supernatant into a clean 2 mL PLG-containing tube (*see* **Notes 5** and **6**).
10. Perform phenol: chloroform DNA extraction as described below in **steps 11–14** (*see* **Note 7**).
11. Add 500 μL of Solution 1 to each tube and mix by inversion.

12. Centrifuge the tubes at 9500 × *g* for 10 min and transfer the supernatant to a clean PLG-containing tube.
13. Add 500 μL of Solution 2 to each tube and invert several times.
14. Spin the tubes 9500 × *g* for 10 min and transfer the upper aqueous layer to a clean 1.5 mL microfuge tube (*see* **Note 8**).
15. Add 800 μL of isopropanol and leave at −20 °C for 1 h.
16. Pellet the DNA by centrifugation at maximum speed (21,000 × *g*) for 5 min. Remove the supernatant.
17. Desalt the DNA by adding 500 μL of 70% ethanol and centrifuge at maximum speed for 5 min. Remove the supernatant.
18. Wash and dehydrate the DNA with 100% ethanol and centrifuge at maximum speed for 5 min. Remove the supernatant (*see* **Note 9**).
19. Invert the microfuge tubes and allow the DNA to dry. Resuspend the pellet in 50–100 μL of AE buffer (Qiagen) or nuclease-free water.
20. Check the final DNA concentration by spectrophotometry (*see* **Note 10** for troubleshooting).

3.2 Quantitative Real-Time PCR

3.2.1 Molecular Beacon Design

Design and synthesis of molecular beacons is described in detail at www.molecular-beacons.org. The probe component of the molecular beacon is the region that is complementary to the target sequence of the amplicon produced by gene-specific primers during PCR. The probes used for the detection of *B. burgdorferi*, *A. phagocytophilum*, and *B. microti* amplicons, as well as for the ACTA1, were all designed to form a stable hybrid with the target at 60 °C, thus allowing for sensitive detection of all pathogen and human DNA in a single multiplex assay. Both the GC content and probe length must be taken into account when designing molecular beacons. At temperatures above the Tm value, the stem melts and results in dissociation of the quencher and fluorophore in the absence of target causing emission of fluorescence. The 5′ and 3′ overhangs of the molecular beacon were designed to form stable hybrids at 5–10 °C above the 60 °C annealing temperature, which facilitates hairpin structure formation. The fluorophores and quenchers for our assay were chosen based on compatibility with a multiplex approach. Biosearch Technologies in California synthesized all probes used in this protocol. Information regarding other companies involved in synthesis can also be found online. Table 2 summarizes the primer pairs and probes utilized in the multiplex approach.

3.2.2 Preparation of B. burgdorferi and Mouse DNA Standards

For the preparation of the standards, uninfected mouse DNA and *B. burgdorferi* genomic DNA must be purified and the concentration determined. The DNA standards are prepared and tested as described below and in Table 3 and Fig. 2.

Table 2
PCR primers and molecular beacon probes

PCR primer/Probe	Sequence[a]	Length (bp)	Tm (°C)	Size of PCR amplicon (bp)	Fluorophore/Quencher
RecF Primer	5′ GTG GAT CTA TTG TAT TAG ATG AGG CTC TCG 3′	30	66	222	FAM/BHQ1
RecR Primer	5′ GCC AAA GTT CTG CAA CAT TAA CAC CTA AAG 3′	30	67		
RecA3 Probe	5′ CTG **GCG GAT ATC CTA GGG GG CGC** CAG 3′	26	68		
5BmTPK Primer	5′ TGA GAG GAA CGA CCA TAG C 3′	19	61	141	CAL Fluor Orange 560/BHQ1
3BmTPK Primer	5′ CCA TCA GGT AAA TCA CAC GAA A 3′	22	62		
BmTPK Probe	5′ CGC GTC **GGT GTT GTT GAC CAG CGG CCG CG** GAC GCG 3′	35	62		
5Aphagocyt Primer	5′ ATG GCT ACT ACG AAG GAT 3′	18	58	152	CAL Fluor Red 610/BHQ2
3Aphagocyt Primer	5′ CGA AGC AAC ATC TCT ACA T 3′	19	58		
Aph1387 Probe	5′ CGG TGC **GAC AAA GAT GCC AGC ACT AAT GCG** GCA CCG 3′	36	62		
5ACTA1 Primer	5′ AGA GCA AGA GAG GTA TCC 3′	18	58	104	Quasar 670/BHQ2
3ACTA1 Primer	5′ CTC GTT GTA GAA GGT GTG 3′	18	58		
ACTA1 Probe	5′ CGC TGC **CCT ATC GAG CAC GGC ATC ATC AC** GCA GCG 3′	35	62		
NidoF Primer	5′ CCA GCC ACA GAA TAC CAT CC 3′	20	70	154	Quasar 670/BHQ2
NidoR Primer	5′ GGA CAT ACT CTG CTG CCA TC 3′	20	70		
Nidogen Probe	5′ CGG CGC **ACC CAG CTT CGG CTC AGT** AGC GCC G 3′	31	77		

[a]Bold font indicates the sequence of the molecular beacon that is complementary to the target

Table 3
Preparation of *B. burgdorferi* and mouse DNA standards

B. burgdorferi DNA	*recA* gene copy number	Mouse DNA	*nidogen* gene copy number
200 ng/5 μL (40 ng/μL)	10^8	200 ng/5 μL (40 ng/μL)	10^5
20 ng/5 μL (4 ng/μL)	10^7	200 ng/5 μL (40 ng/μL)	10^5
2 ng/5 μL (0.4 ng/μL)	10^6	200 ng/5 μL (40 ng/μL)	10^5
200 pg/5 μL (40 pg/μL)	10^5	200 ng/5 μL (40 ng/μL)	10^5
20 pg/5 μL (4 pg/μL)	10^4	200 ng/5 μL (40 ng/μL)	10^5
2 pg/5 μL (0.4 pg/μL)	10^3	200 ng/5 μL (40 ng/μL)	10^5
200 fg/5 μL (40 fg/μL)	10^2	200 ng/5 μL (40 ng/μL)	10^5
20 fg/5 μL (4 fg/μL)	10	200 ng/5 μL (40 ng/μL)	10^5
2 fg/5 μL (0.4 fg/μL)	1	200 ng/5 μL (40 ng/μL)	10^5

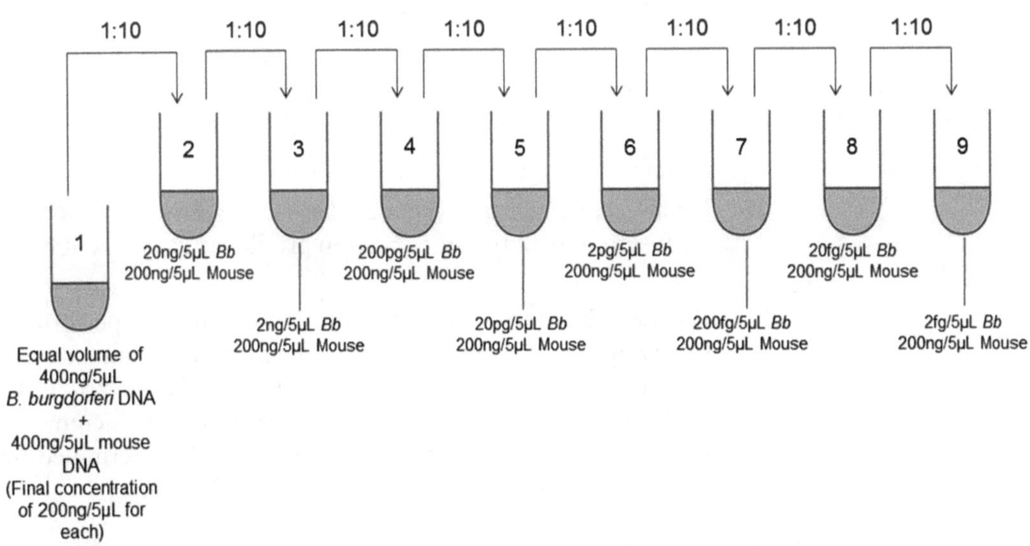

Fig. 2 Schematic for *B. burgdorferi* and mouse DNA standard preparation. Mixing equal volumes of 200 ng/5 μL *B. burgdorferi* and mouse DNA yields a stock containing 10^8 copies of the *recA* gene and 10^5 copies of the *nidogen* gene. Serial dilutions using 200 ng/5 μL of mouse DNA as diluent allows maintenance of a uniform copy number of the *nidogen* gene (10^5/5 μL) while achieving tenfold dilutions of the *recA* copy number

1. Prepare 100 μL of 80 ng/μL stock of both *B. burgdorferi* and mouse genomic DNA.
2. Combine 100 μL of 80 ng/μL *B. burgdorferi* DNA with equal volume of mouse DNA stock solution. Based on the size of the respective genome this corresponds to 10^8 copies of the *recA* gene and 10^5 copies of the *nidogen* gene (*see* **Note 11**).
3. Prepare 1 mL of 40 ng/μL mouse DNA stock as diluent to maintain the *nidogen* at 10^5 copies for all dilutions.
4. Vortex the mixture prepared in **step 2**, with 10^8 *B. burgdorferi recA* copy number and 10^5 copies of *nidogen*, briefly centrifuge and perform 1:10 serial dilutions in mouse DNA diluents prepared in **step 3** to a final dilution to obtain one copy of *B. burgdorferi recA*.

Generating a linear standard curve ($R^2 \geq 0.9$) is necessary for the accurate interpretation of all rtPCR data (Fig. 3). Individual monoplex assays can be used to confirm the accuracy of newly prepared standards, prior to utilizing the standards for multiplex analysis.

3.2.3 Real-Time PCR Assays

All our rtPCR assays are performed using a Bio-Rad CFX96 Touch Real-time PCR system even though other available systems can also be used that can detect four or more fluorophores. All assays utilize 0.2 mL optical tubes or semi-skirted 96-well PCR plates depending on the number of samples. Reagents, master mix preparation, and conditions for rtPCR assays are described in Table 4.

1. Add 20 μL of master mix into each well.
2. Subsequently add 5 μL of standard or template DNA for a final reaction volume of 25 μL (*see* **Note 12**).
3. Seal plate with a transparent, non-fluorescing silicone film adhesive or cap tubes tightly with flat caps.
4. Briefly centrifuge tubes/plate. Place tubes into the real-time PCR system with the amplification profile setup as described in Table 4.
5. For *Borrelia* species identification, a melt curve is performed following the PCR by decreasing the temperature immediately after the final polymerization step from 72 °C to 25 °C to allow annealing. Increase the temperature at 1 °C increments from 25 °C to 95 °C for 2 min. Tm values are determined by monitoring fluorescence levels (*see* **Note 13**).

3.3 Testing of Samples

Storage of B. burgdorferi spiked blood samples at 4 °C does not affect detection sensitivity.

Often there is a lag period between collection of patient samples at the clinics and their arrival at the testing laboratory. Therefore, it is important to know if the detection sensitivity of *B.*

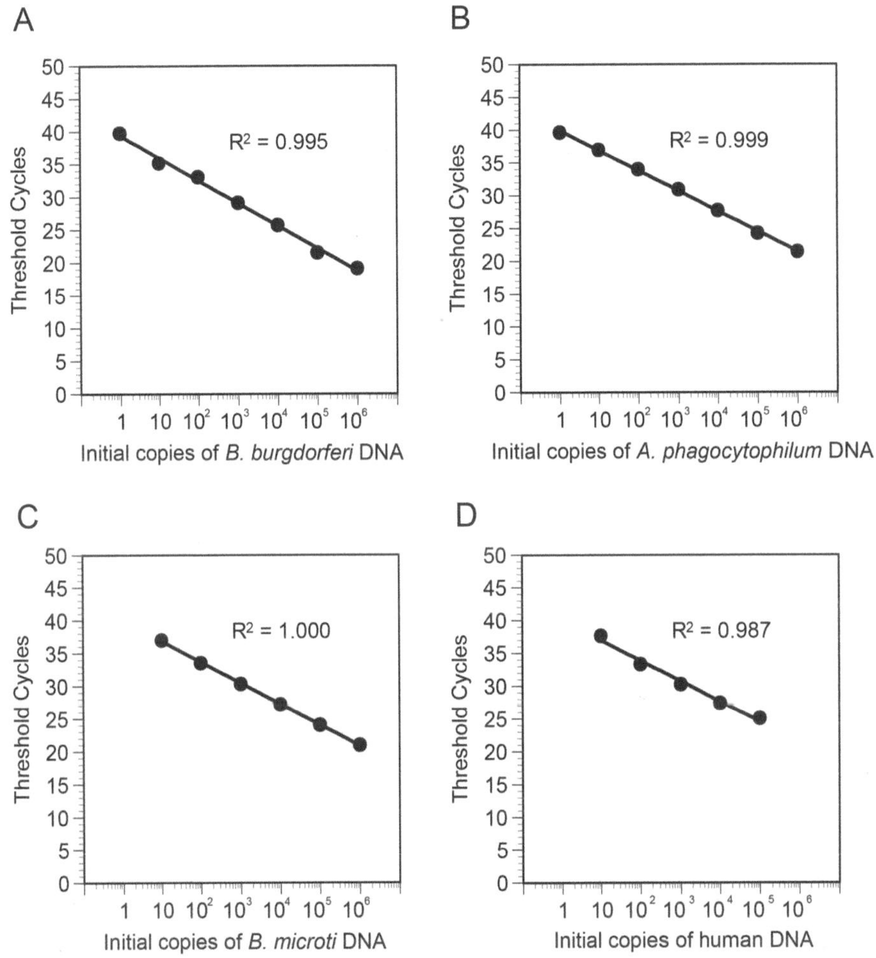

Fig. 3 High coefficient of correlation of *B. burgdorferi, A. phagocytophilum*, and *B. microti* DNA diluted in fixed (10^5 copies) of the human gene encoding alpha actin one protein, ACTA1 indicate efficiency of our quadruplex rtPCR. (**a–c**). Serial tenfold dilutions of all three pathogens together were prepared in fixed human DNA concentration (10^5 copies of human gene ACTA1) and quadruplex assays were conducted. High sensitivity of detection of all three pathogens was maintained as indicated by high coefficient of correlation between genes copy number and amplification cycle number. Overlapping peaks of human ACTA1 indicated the sensitivity of detection was not affected for human gene even in quadruplex assay (*data not shown*). (**d**). Serial dilution of human DNA in a monoplex assay indicates that the rtPCR can detect human ACTA1 in an efficient manner until a copy number of 10^5 per reaction. Therefore, 10^5 copy number of ACTA1 was used for duplex and quadruplex assays

burgdorferi, and other tick-borne pathogens are adversely affected by storage. We spiked 20 mL of human blood with *B. burgdorferi* (100 spirochetes/mL of blood) and prepared aliquots, which were stored at 4 °C in a refrigerator to reflect the clinical situation of patient sample collection and storage. At 48 h intervals two aliquots were removed and DNA isolated for rtPCR. Interestingly, there was no significant change in the detection of *B. burgdorferi* in blood until 2 weeks of storage (Fig. 4). We anticipate that patient

Table 4
Master mix setup and thermal profile for real-time PCR assays

Master mix setup for duplex PCR assay			Master mix setup for quadruplex PCR assay		
Reagent	**Final concentration**	**Volume (μL)**	**Reagent**	**Final concentration**	**Volume (μL)**
10× AmpliTaq Gold buffer	1×	2.5	10× AmpliTaq Gold buffer	1×	2.5
BSA (5 mg/mL)	0.5 mg/mL	2.5	BSA (5 mg/mL)	0.5 mg/mL	2.5
$MgCl_2$ (25 mM)	3 mM	3	$MgCl_2$ (25 mM)	3 mM	3
dNTPs (25 mM)	0.25 mM	0.25	dNTPs (25 mM)	0.25 mM	0.25
dH2O		4.25	dH2O		6.75
RecF primer (5 μM)	0.5 μM	2.5	RecF primer (50 μM)	0.5 μM	0.25
RecR primer (5 μM)	0.5 μM	2.5	RecR primer (50 μM)	0.5 μM	0.25
RecA3 probe (100 ng/μL)	50 ng	0.5	RecA3 probe (100 ng/μL)	25 ng	0.25
NidoF primer (5 μM)	0.1 μM	0.5	5BmTPK primer (50 μM)	0.5 μM	0.25
NidoR primer (5 μM)	0.1 μM	0.5	3BmTPK primer (50 μM)	0.5 μM	0.25
Nidogen probe (100 ng/μL)	50 ng	0.5	BmTPK probe (100 ng/μL)	25 ng	0.25
–			5Aphagocyt (50 μM)	0.5 μM	0. 25
–			3Aphagocyt (50 μM)	0.5 μM	0.25
–			Aph1387 probe (100 ng/μL)	25 ng	0.25
–			5ACTA1 (5 μM)	0.1 μM	0.5
–			3ACTA1 (5 μM)	0.1 μM	0.5
–			ACTA1 probe (100 ng/μL)	25 ng	0.25
AmpliTaq Gold (5 U/μL)	2.5 U	0.5	AmpliTaq Gold (5 U/μL)	7.5 U	1.5
Master mix volume		**20**	**Master mix volume**		**20**
Template		**5**	**Template**		**5**
Total volume		**25**	**Total volume**		**25**

Thermal Profile for qPCR assays

95 °C for 5 min
50 cycles of the following:
95 °C for 15 s
60 °C for 30 s
72 °C for 20 s

Bold letters indicate the probe sequence complementary to Bb-RecA3 target sequence

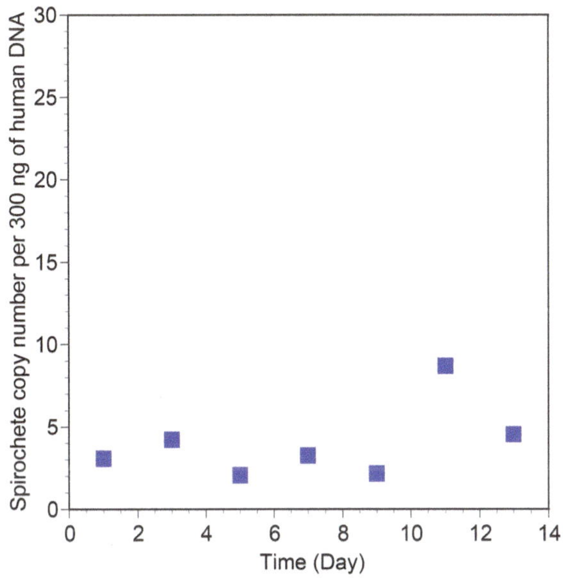

Fig. 4 Sensitivity of detection of *B. burgdorferi* is not affected by storage of blood in refrigerator. Human blood samples were collected and spiked with *B. burgdorferi* and 1 mL aliquots prepared. Samples were then incubated at 4 °C for up to 14 days. At 48 h intervals two aliquots were used for DNA isolation and rtPCR performed. Storage of up to 14 days showed no significant effect on the detection efficiency of the assay for *B. burgdorferi*

samples are usually tested within a week of collection. Therefore, our rtPCR assay will allow time for proper storage and transport within 2 weeks without affecting the efficiency of detection for *B. burgdorferi*, and likely other pathogens.

Although we have optimized conditions and presented data using blood samples, it is highly likely that this assay will be equally successful using cerebrospinal fluid, synovial fluid, or even urine samples from patients. We expect that urine may be a less invasive alternative to blood samples for the detection of *B. burgdorferi*. Furthermore, during animal infection with *B. burgdorferi*, we typically find prominent colonization of the bladder by *B. burgdorferi*. Therefore, clinical application of the rtPCR assay we have outlined has the potential to allow for more sensitive diagnosis of tick-borne pathogens in blood and Lyme spirochetes in other patient samples.

4 Notes

1. For DNA recovery, add extract directly into the center of the DNA purification column membrane, wait 5 min, and centrifuge at 9500 × g for 1 min to achieve the highest recovery. We have found that repetition of this step using the same column improves DNA yield.

2. If DNA quality or concentration is poor, return to **step 9** (Subheadings 3.1.1 and 3.1.2) and perform a second phenol: chloroform extraction.
3. This protocol is suitable for all mouse tissues except for the heart. In the case of the heart to expedite the process, it is necessary to first finely chop the organ in small pieces. A small petri dish can be used for this purpose and transfer chopped heart to a 15 mL tube. Add 2.5 mL of 0.1% collagenase A and incubate at 37 °C for 4 h with intermittent vortexing to homogenize the tissue.
4. A shaking water bath set to 60 °C can eliminate the need for multiple rounds of vortexing. Tissues can be incubated in a shaking water bath for approximately 24 h, while monitoring the amount of homogenization over the incubating period.
5. PLG-containing MaXtract tubes (Qiagen) allow clean separation of the phenol: chloroform phase and aqueous phase by a PLG solid layer preventing the carryover of contaminants and yielding cleaner DNA. Prior to the addition of homogenate the high-density gel must be pelleted by centrifugation.
6. Be careful to avoid any mouse undigested tissue. This may require multiple rounds of centrifugation and supernatant collection.
7. Phenol: chloroform solutions should be prepared immediately before use.
8. Occasionally, the tissues may have debris or gel carryover that will interfere with DNA isolation. If this is the case, remove as much supernatant as possible without carrying over the debris to a clean microfuge tube and then continue.
9. At final stage, the DNA pellet should be translucent white. If it not, or if the debris remains, gently resuspend and transfer top of the supernatant to a clean tube carefully while avoiding the debris and repeat DNA recovery.
10. There are several ways to troubleshoot the problem of low DNA concentration or its poor quality, such as: (A) centrifugation of the samples to pellet the debris followed by careful transfer of the supernatant to a clean microfuge tube avoiding debris pellet, (B) repeating ethanol precipitation step with clear top supernatant of suspended DNA, (C) a second round of phenol: chloroform extraction to remove protein impurities, (D) a final pass of the DNA through the Qiagen DNeasy Blood and Tissue kit columns can improve DNA quality [15].
11. The size of the *B. burgdorferi* genome is ~1.5 Mb, so in 2 ng of DNA there are 10^6 copies of *recA* gene. The mouse genome is ~2 Gb with one copy of *nidogen* gene on each chromosome. Thus, there are 10^5 copies of *nidogen* gene in 200 ng of DNA. These calculations allow for quantification of bacterial load relative to quantity of mouse DNA.

12. When preparing to add template and standards to PCR tubes/plates, make sure each sample is very well mixed by vortexing thoroughly, followed by pipetting the entire volume up and down at least ten times before drawing up 5 μL. When adding the standards, adding from the lowest concentration to the highest allows good standard curve preparation.
13. Tm values can vary depending on reaction conditions (i.e., DNA template, $MgCl_2$ concentrations, primer concentrations, etc.). Additionally, Tm values can vary between experimenter and thermal cycler. Therefore, it is imperative that the same reaction conditions used for the multiplex rtPCR assay be maintained for *Borrelia* speciation, in order to minimize variance.

Acknowledgment

This work was supported by NIH grant AI089921, and New Jersey Health Foundation and The Office of Research & Economic Development grants to NP.

References

1. Marques AR (2010) Lyme disease: a review. Curr Allergy Asthma Rep 10:13–20
2. Wright WF, Riedel DJ, Talwani R, Gilliam BL (2012) Diagnosis and management of Lyme disease. Am Fam Physician 85:1086–1093
3. Steere AC (2001) Lyme disease. N Engl J Med 345:115–125
4. Dantas-Torres F, Chomel BB, Otranto D (2012) Ticks and tick-borne diseases: a one health perspective. Trends Parasitol 28:437–446
5. Ginsberg HS (2008) Potential effects of mixed infections in ticks on transmission dynamics of pathogens: comparative analysis of published records. Exp Appl Acarol 46:29–41
6. Belongia EA (2002) Epidemiology and impact of coinfections acquired from Ixodes ticks. Vector Borne Zoonotic Dis 2:265–273
7. Wormser GP, Dattwyler RJ, Shapiro ED, Halperin JJ, Steere AC, Klempner MS, Krause PJ, Bakken JS, Strle F, Stanek G, Bockenstedt L, Fish D, Dumler JS, Nadelman RB (2006) The clinical assessment, treatment, and prevention of Lyme disease, human granulocytic anaplasmosis, and babesiosis: clinical practice guidelines by the infectious diseases society of America. Clin Infect Dis 43:1089–1134
8. Maurin M (2012) Real-time PCR as a diagnostic tool for bacterial diseases. Expert Rev Mol Diagn 12:731–754
9. Wilson ML (2011) Recent advances in the laboratory detection of mycobacterium tuberculosis complex and drug resistance. Clin Infect Dis 52:1350–1355
10. Lee JH, Uhl JR, Cockerill FR 3rd, Weaver AL, Orvidas LJ (2008) Real-time PCR vs standard culture detection of group a beta-hemolytic streptococci at various anatomic sites in tonsillectomy patients. Arch Otolaryngol Head Neck Surg 134:1177–1181
11. Green BN, Johnson CD, Egan JT, Rosenthal M, Griffith EA, Evans MW (2012) Methicillin-resistant *Staphylococcus aureus*: an overview for manual therapists(). J Chiropr Med 11:64–76
12. Selva L, Esteva C, Gene A, de Sevilla MF, Hernandez-Bou S, Munoz-Almagro C (2010) Direct detection of Streptococcus Pneumoniae in positive blood cultures by real-time polymerase chain reaction. Diagn Microbiol Infect Dis 66:204–206
13. Colborn JM, Kosoy MY, Motin VL, Telepnev MV, Valbuena G, Myint KS, Fofanov Y, Putonti C, Feng C, Peruski L (2010) Improved detection of bartonella DNA in mammalian hosts and arthropod vectors by real-time PCR using the NADH dehydrogenase gamma subunit (nuoG). J Clin Microbiol 48:4630–4633
14. Riehm JM, Rahalison L, Scholz HC, Thoma B, Pfeffer M, Razanakoto LM, Al Dahouk S, Neubauer H, Tomaso H (2011) Detection of

Yersinia Pestis using real-time PCR in patients with suspected bubonic plague. Mol Cell Probes 25:8–12

15. Saidac DS, Marras SA, Parveen N (2009) Detection and quantification of Lyme spirochetes using sensitive and specific molecular beacon probes. BMC Microbiol 9:43
16. Chan K, Marras SA, Parveen N (2013) Sensitive multiplex PCR assay to differentiate Lyme spirochetes and emerging pathogens Anaplasma phagocytophilum and Babesia microti. BMC Microbiol 13:295
17. Marras SAE, Russell Kramer F, Tyagi S (1999) Multiplex detection of single-nucleotide variations using molecular beacons. Genet Anal 14:151–156
18. Tyagi S, Bratu DP, Kramer FR (1998) Multicolor molecular beacons for allele discrimination. Nat Biotechnol 16:49–53
19. Marras SA, Kramer FR, Tyagi S (2002) Efficiencies of fluorescence resonance energy transfer and contact-mediated quenching in oligonucleotide probes. Nucleic Acids Res 30:e122
20. Liveris D, Schwartz I, McKenna D, Nowakowski J, Nadelman R, Demarco J, Iyer R, Bittker S, Cooper D, Holmgren D, Wormser GP (2012) Comparison of five diagnostic modalities for direct detection of Borrelia burgdorferi in patients with early Lyme disease. Diagn Microbiol Infect Dis 73:243–245
21. Aguero-Rosenfeld ME, Wang G, Schwartz I, Wormser GP (2005) Diagnosis of Lyme borreliosis. Clin Microbiol Rev 18:484–509
22. Schotthoefer AM, Frost HM (2015) Ecology and epidemiology of Lyme borreliosis. Clin Lab Med 35:723–743
23. Herwaldt BL, Linden JV, Bosserman E, Young C, Olkowska D, Wilson M (2011) Transfusion-associated babesiosis in the United States: a description of cases. Ann Intern Med 155:509–519

Chapter 11

Methods for Real-Time PCR-Based Diagnosis of *Chlamydia pneumoniae*, *Chlamydia psittaci*, and *Chlamydia abortus* Infections in an Opened Molecular Diagnostic Platform

Onya Opota, René Brouillet, Gilbert Greub, and Katia Jaton

Abstract

The advances in molecular biology of the last decades have dramatically improved the field of diagnostic bacteriology. In particular, PCR-based technologies have impacted the diagnosis of infections caused by obligate intracellular bacteria such as pathogens from the *Chlamydiacae* family. Here, we describe a real-time PCR-based method using the Taqman technology for the diagnosis of *Chlamydia pneumoniae*, *Chlamydia psittaci*, and *Chlamydia abortus* infection. The method presented here can be applied to various clinical samples and can be adapted on opened molecular diagnostic platforms.

Key words *Chlamydia pneumoniae*, *Chlamydia psittaci*, *Chlamydia abortus*, Molecular diagnostic, Real-time PCR, DNA extraction, Taqman

1 Introduction

Chlamydiacae are obligate intracellular bacteria, among which several species are pathogenic toward humans and can cause a broad range of diseases. *Chlamydia trachomatis* is involved in urogenital infection as well as ocular, joint, and oropharyngeal infections. *Chlamydia pneumoniae* and *Chlamydia psittaci* are primarily associated with community-acquired pneumonia (CAP). *C. pneumoniae* have been essentially associated with infections in humans [1], but some studies suggest an association with other mammals such as koalas [2–8]. *C. pneumoniae* might also be a causative agent of an asthma-like syndrome in children [9]. *C. psittaci* is the etiologic agent of a respiratory zoonosis transmitted by birds, in particular parrots and parakeets, but also chicken from the food industries as well as feral pigeons [2]. *C. abortus* is genetically closely related to *C. psittaci* but displays a distinct animal and tissue tropism as it can colonize the placenta of cows, goats, cattles, pigs, and horses, which can lead to abortion. *C. abortus* can be

Kimberly A. Bishop-Lilly (ed.), *Diagnostic Bacteriology: Methods and Protocols*, Methods in Molecular Biology, vol. 1616, DOI 10.1007/978-1-4939-7037-7_11, © Springer Science+Business Media LLC 2017

transmitted to humans, for instance through the exposure to infected animal abortive tissues, with the same outcome for pregnant women [2].

The diagnosis of infections due to intracellular bacteria has been dramatically improved by PCR-based methods developed in the last decades. This is particularly true for the diagnosis of *C. trachomatis*, for which several commercial systems have been developed. The laboratory of molecular diagnostic of the Institute of Microbiology of the Lausanne University Hospital (Lausanne, Switzerland), developed a molecular diagnostic platform recently described by Greub and colleagues, which allows one to perform multiple PCRs simultaneously targeting different pathogens, thanks to the standardization of the parameters (i.e., amplicon length, and primers and probes T_m) of the reactions [10]. This molecular platform based on the Taqman-probes technology (Applied Biosystems) allows for performance of up to 91 different PCR reactions corresponding to 69 pathogens and/or resistance genes in a single microplate [10]. Among them are a real-time PCR for the detection of *C. pneumoniae* [11] and a real-time duplex PCR for the detection of *C. psittaci* and *C. abortus* infections from various clinical samples (Table 1), both based on the Taqman probe technology [12]. The *C. pneumoniae* PCR is based on the *pst1* gene and the *C. psittaci*-*C. abortus* duplex PCR was designed as follows: (1) PCR1 targets a DNA sequence of the 16S–23S rRNA operon allowing the detection of both *C. psittaci* and *C. abortus* and (2) PCR 2 targets the coding DNA sequence CPSIT_0607 so far unique to *C. psittaci* [12]. In this chapter, the methods to achieve the *C. pneumoniae* PCR and the *C. psittaci*-*C. abortus* duplex PCR on an opened molecular diagnostic platform are detailed. For this specific chapter, part of the samples' processing methods and infrastructure descriptions have been extracted from the accredited documentation of the Laboratory of Diagnostic of the Institute of Microbiology of the University of Lausanne.

2 Materials

2.1 Laboratory Organization

All the procedures should be carried out according to molecular diagnostic principles aimed to avoid contaminations with microorganisms or nucleic acids (*see* **Note 1**).

2.2 Sample Processing

1. *N*-Acetyl cysteine solution: in a 50 ml conical tube, containing 1.0 g of *N*-acetyl cysteine, add 50 ml of Tris-Sodium-Citrate-di-hydrate buffer. The solution should be used the same day.
2. Molecular biology grade PBS: DNase-Free, RNase-Free, does not contain detectable amounts of nucleic acid or any nucleic-acid extractions' compatible molecular biology grade solution.

Table 1

Primers and probes

Pathogen	Oligonucleotide name	Target gene	Sequence (5′–3′)	Modification, fluorochrome	Amplicon length	Final concentration (μM)	References
C. pneumoniae	CPTM1	*pst1* gene	CATGGTGTCATTCGCCAAGT	–		0.2	Welti et al. [11]
	CPTM2	*pst1* gene	CGTGTCGTCCAGCCATTT TA	–		0.2	Welti et al. [11]
	CP	*pst1* gene	TCTACGTTGCCTCTAAGAGAAAA CTTCAAGTTGGA	3′-VIC, 5′-TAMRA	82	0.1	Welti et al. [11]
C. psittaci and *C. abortus*	CPSI_F	16S–23S operon	AAGGAGAGAGGCGCCCAA	–		0.35	Opota et al. [12]
	CPSI_R_LNA	16S–23S operon	CAA[C]CTAGTCAAACCGTCCTAA	LNA		0.35	Opota et al. [12]
	CPSI_P_MGB	16S–23S operon	ACTGGGATGAAGTCGTAAC	FAM, DQ	133	0.2	Opota et al. [12]
C. psittaci	CPSI_00F	CDS CPSI_0607	AGCATTAGCCAGCGCTTTAGA	–		0.35	Opota et al. [12]
	CPSI_00R_147C/G	CDS CPSI_0607	TCTCTGAGCAAAAC/GACTGCGT	–		0.35	Opota et al. [12]
	CPSI_00P_MGB	CDS CPSI_0607	ACAAAGACCTGGCGAGTA	VIC, DQ	118	0.2	Opota et al. [12]

LNA locked nucleic acid, *MGB* minor groove binder, *FAM* 6-carboxy-fluorescein, *VIC* TaqMan VIC reporter dye, *DQ* dark quencher, *BHQ* Black hole quencher

2.3 Material for DNA Extraction

1. Several automated instruments exist for DNA extraction, in the Institute of microbiology, nucleic-acid extraction is achieved with the MagNA Pure 96® instrument (Roche): an automated system for the extraction of nucleic-acid from bacteria (as well as other microorganisms such as viruses) using a technology based on magnetic glass particles. The MagNA Pure 96® instrument reagents are those provided by the manufacturer. This nucleic-acid extraction instrument has been associated with a liquid-handling distribution system, the STARlet® instrument (Hamilton®) [13] (*see* **Note 2**).
2. Molecular biology grade PBS: DNase-Free, RNase-Free, does not contain detectable amounts of nucleic acid or any nucleic-acid extractions' compatible molecular biology grade solution.

2.4 Assembly of the PCR Plate and Amplification

1. TaqMan Universal Master Mix (Applied Biosystems).
2. Primers and probes (Table 1).
3. Molecular biology grade water: DNase- and RNase-free water, which does not contain detectable amounts of nucleic acid.
4. Positive control. The positive controls consist of synthetic plasmids containing the exact PCR amplicon [11, 12].
5. PCR instrument (*see* **Notes 3** and **4**).

3 Methods

3.1 Sample Processing

In clinical practice, the detection of *C. pneumoniae, C. psittaci*, and *C. abortus* can be achieved from a broad range of samples (Table 2). Depending on the nature of the sample (liquid, viscous, or solid), a specific processing will be required either to concentrate (liquid samples) or to homogenize (viscous or solid samples) the specimen.

1. Introduce the lab request into the laboratory information system (LIS) to generate the barcode labels necessary to manage tubes during the analytic process (*see* **Note 5**).
2. For each sample, label three screw cap tubes (2 ml) with the barcode as follows: one "native" tube, one "native-aliquot" tube, and one "DNA tube."
3. Under the laminar flow dedicated for molecular diagnostics in the "specimen-receiving laboratory," distribute the sample into the native-aliquot tube and the native tube. The native tube will be frozen at −80 °C and kept as backup in case of need. If necessary several native tubes can be stored; a suggestion for biopsies would be to store both the native samples and the processed samples in distinct tubes. The aliquot-native tube together with the DNA tube will be transported to the

molecular diagnostic laboratory and used for nucleic acids extraction. Viscous samples such as respiratory secretions should be homogenized by liquefaction whereas solid samples such as biopsies and fragments should be crushed as described below.

4. Liquefaction of viscous secretions: respiratory secretions (sputum, bronchial aspirate, stomach tube) for the detection of *C. pneumoniae* and *C. psittaci* can be too viscous to be homogenized by simple vortexing. Liquefaction can be achieved using *N*-acetyl-L-cysteine, a reducing substance having free thiol groups (–SH), to reduce specimen viscosity by breaking the disulfide bonds of glycoproteins that constitute the bronchial mucus. Any other viscous fluids shall also be liquefied. To do so, visually control the viscosity of the sample; if liquefaction is necessary, transfer a suitable amount of the sample in a 15 or 50 ml conical tube using a single use plastic Pasteur pipette, under the hood with laminar flow. Add an equal amount of *N*-acetyl-L-cysteine solution. A larger amount of *N*-acetyl-L-cysteine solution can be needed for very sticky samples. Vortex and visually check the liquefaction of the sample and if necessary, leave the sample on a rotor for up to 30 min. At the end of the incubation, visually check the liquefaction of the sample and centrifuge the sample for 30 min at 3000 × *g* and then remove the supernatant using a pipette with filter tips, leaving 1–2 mL of liquid in which the pellet will be resuspended. Transfer the amount necessary for nucleic acids extraction (more than 200 μl) into the tube labeled as "aliquot-native" (*see* **Note 6**), and keep the rest in the tube labeled as "native." Store the native tube at –80 °C in the core specimen receiving laboratory and transport the aliquot-native tube and the DNA tube to the molecular diagnostic laboratory.
5. Crushing of solid specimens. Solid samples such as biopsies and fragments that cannot be homogenized either using the vortex or by liquefaction should be crushed as follows. Under the hood with laminar flow put the little pieces in the crushing device with part of the solution of the native sample (if any) or add a solution adapted to the DNA extractor (PBS) using a pipette with filtered tips (*see* **Note 7**). If the specimen is too big, take a piece of it with sterile forceps and place it in a sterile Petri dish to cut it into several small pieces. Crush a piece of the specimen and keep the rest in reserve. Start the crushing device. Generally, the fragments are successfully crushed, if not, this procedure is still sufficient to release microorganisms from the specimen by compression. In this case, avoiding some remaining large fragments, transfer the crushed sample (volume) in the aliquot-native tube, and the rest in the native tube for storage at –20 °C. If necessary add some molecular biology

Table 2

Clinical specimens and pathogens generally used for the diagnosis of *Chlamydia* infections

	Sample type	*C. pneumoniae*	*C. psittaci*	*C. trachomatis*	*C. abortus*
Oral	Mouth swab			×	
Respiratory specimen	Nasal swab	×	×		
Respiratory specimen	Nasopharyngeal secretions	×	×		
Respiratory specimen	Sputum	×	×		
Respiratory specimen	Bronchial secretion	×	×		
Respiratory specimen	Bronchoalveolar lavage	×	×		
Urogenital specimen	Uterus and urethral smear			×	×
Urogenital specimen	Urethral swab			×	
Urogenital specimen	Prostate biopsies			×	
Urogenital specimen	Fragment of placenta			×	×
Anal	Anal swab			×	
Osteo-articular	Joint fluid			×	
Osteo-articular	Prosthetic fragment			×	
Vascular	Drain fluid from aortic valve	×	×		
Animal specimen	Bird spleen		×		
Animal specimen	Birds choanal or cloacal swabs	×			
Animal specimen	Fragment of sheep placenta				×
Non-comprehensive list					

grade water into the aliquot-native tube to reach the minimum volume suitable for nucleic acids extraction. Vortex the tube. Transport the aliquot-native tube and the DNA tube to the molecular diagnostic laboratory.

3.2 DNA Extraction

1. Using the STARlet® liquid-handling instrument, transfer 200 µl of the sample tube in the extraction plate of the MagNA Pure 96® instrument.
2. Extraction control (EC). The EC corresponds to a tube submitted to the same extraction protocol that the clinical

specimen but in which the volume of the clinical sample is replaced by an equal volume of a solution adapted to the DNA extractor (PBS). An extraction control is needed for each run of extraction (*see* **Note 8**). The extraction control must be negative when the PCR specific for the pathogen(s) tested is(are) made.

3. Transfer the 96-well microplate into the MagNAPure 96® instrument and start the nucleic acids extraction program according to the user manual. Nucleic acids can be eluted either in 50 or 100 μl of elution buffer depending on the chosen program and are maintained at 4 °C (*see* **Note 9**).

3.3 Preparation of the PCR Controls

1. Negative control of PCR. Each run of amplification should contain a PCR negative control that consists of the extraction control used as a template. The negative control is used to test the reactivity of the component of the reaction mixture. It should not be contaminated with target DNA and should not allow nonspecific amplification.
2. Positive control. Positive controls consist of synthetic plasmid DNA containing the target sequence of the PCRs [11, 12] (*see* **Note 10**). For each run of PCR, three reactions with three dilutions of the positive controls containing 10, 10^2, and 10^3 DNA copies per reaction should be added. They will serve both to generate the standard curve that will be used for the quantification of the positive control and to determine the sensitivity of the reaction based on the positive amplification, the reaction containing ten copies of DNA.
3. Inhibition control. The presence of PCR inhibitors should be tested for each sample. To do so, the inhibition control reaction consists of a reaction in which 200 copies of the positive control are added to the reaction mixture containing the DNA specimen to be tested [10].

3.4 Preparation of the PCR Mix, Design, and Assembly of the PCR Microplate and Amplification

1. Design of the Taqman PCR plate. This can be achieved using dedicated software such as the SDS 2.4.1 software (Applied Biosystems) allowing the design of either 96- or 384-well PCR plates [10]. The same final volume (20 μl) is convenient for both types of plate. It is strongly recommended to do each analysis in duplicate (or even in triplicate). When an internal control is not used, an additional well for the inhibition control is needed, if not using internal inhibition control. Moreover, there is a need for at least one well for the inhibition control, one well for the negative control and three wells for the standard curve. A standard curve is required for each PCR. One regression curve is required for the PCR1 and another for the PCR2 [12].

2. In the DNA-free laboratory, prepare the PCR mix. A single mix is needed for the detection of *C. psittaci* and *C. abortus*. In a final volume of 20 μl, add 5 μl of the extracted DNA, the forward and reverse primers and the probes at the final concentration indicated in Table 2. The inhibition control reaction consists of the same reaction in which, 200 copies of the control plasmids are added. As an extraction negative control the DNA is replaced by the same volume of the extraction control. As a negative control of the PCR, the DNA is replaced by molecular grade water.
3. Assembly of the PCR plate. The assembly of the PCR plate can be done manually for 96-well plates (*see* **Note 11**) or using automated instruments for 384-well plates [10]. The reactions are achieved in a final volume of 20 μl with 5 μl of DNA sample [11, 12].
4. Amplification. Run the ABI 7900 instrument or similar thermocycler using the following cycling conditions: 2 min at 50 °C, 10 min at 95 °C followed by 45 cycles of 15 s at 95 °C, and 1 min sec at 60 °C.

3.5 Interpretation of the Results

The SDS 2.4.1 software is used for the analysis and interpretation of the results.

The results (qualitative and quantitative) are then checked and introduced into the LIS system by biomedical technicians (technical validation).

At the end of the process, the final results are validated by a clinical microbiologist (biomedical validation).

1. Analyze the results of the positive controls with adequate software [10].
2. Internal quality control. For an analysis to be valid: (a) the run should pass the internal quality control and (b) the positive controls of ten copies must be detected.
3. Negative controls. Control that there is no amplification in the negative control reaction.
4. Inhibition control. Control that there is amplification in the inhibition control tube containing 200 copies of the positive control plasmid (at least of 50 copies).
5. Positive result. A result is positive if the fluorescence reaches the threshold automatically set by the software or manually set by the user according to the instrument.
6. Interpretation of the *C. psittaci*-*C. abortus* PCR (*see* **Note 12**).
7. If you suspect a contamination, when for instance only one reaction out of three reactions is positive or when amplification occurs in the extraction control or in the PCR negative control, you can follow the procedure described by Greub and colleagues for these situations [10].

4 Notes

1. As described by Greub and colleagues, it is important to organize the laboratory of molecular diagnostic in different rooms/spaces corresponding to pre-amplification and post-amplification areas [10]. Moreover, the processing of clinical samples in the laboratory where the samples are received should be achieved under a laminar flow dedicated for molecular diagnostics that is distinct from laminar flows dedicated for conventional culture-based microbiology diagnostics. Sterile samples should also be processed in dedicated laminar flows. The use of disposable lab coats, gloves, and pipettes with filter tips is recommended. Thus, we recommend the following infrastructures and instruments: (a) the sample reception laboratory should be equipped with a hood with laminar flow and UV, a vortex, a crushing instrument, a centrifuge, disposable lab coats and gloves, pipettes and filter tips, 2 ml screw cap micro-tubes, sterile forceps, and sterile Petri dishes; (b) the nucleic acids' extraction laboratory should also be equipped with a hood with laminar flow and UV, a vortex, pipettes and filter tips, disposable lab coats and gloves without mineral powders such as talc to prevent any deposition in the extraction tubes; (c) the PCR master-mix should be prepared in a DNA-free laboratory equipped with a hood with UV but without laminar flow; for all the post extraction area, hood with laminar flow should be prohibited in order to avoid the deposition of nucleic acids (NA) in opened tested tubes. In addition, the following instruments should be available in this laboratory: a vortex, a micro-centrifuge, micro-tubes (0.2, 0.7, and 1.5 ml), 2 ml screwed tube, pipettes and filter tips (0–10, 2–20, 20–200 et 200–1000 μl); (d) the positive control should be prepared and stored in a "DNA laboratory" equipped with a hood with UV but without laminar flow, a vortex, a micro-centrifuge, micro-tubes, pipettes, and filter tips; (e) It is recommended to assemble the PCR plate a dedicated pre-amplification laboratory different from the positive control laboratory.
2. If processing a large number of samples, it is recommended to select automated distribution systems that can recognize bar-coded tubes.
3. The assembly of the PCR plate can be achieved manually for 96-well plates or alternatively can be done by automated instruments to increase the precision and to avoid mistakes, especially when preparing 384-well plates [10].
4. The PCR conditions described herein for the detection of *C. pneumoniae*, *C. psittaci*, and *C. abortus* have been optimized

for the PCR instrument ABI 7900 (Applied Biosystems) [12]. The reagent concentrations and the amplification program should be adapted if using other instruments [10].

5. Tubes management is crucial all along the analytic process to avoid major errors such as inversions or contaminations. We strongly recommend the use of barcoded tubes.
6. If using an automated system such as the STARlet® instrument (Hamilton®), do not forget to provide the dead volume of the device.
7. To avoid the dilution of the specimen, which could negatively impact the sensitivity of the PCR, do not add too much solution for the crushing of the sample.
8. We recommend placing the extraction control at the end of the series where contamination is more expected to occur rather than at the first position of a series.
9. It is recommended to store the DNA sample at −80 °C.
10. The concentration (DNA copies per ml) of the positive control should be precisely determined. The stock solution of the positive control should be stored in a separate, dedicated room.
11. The assembly of the PCR plate can be achieved manually especially for 96-well plates. In the DNA-free laboratory prepare the PCR mix without the DNA. Place a 96-well micro-plate on a chilled metal rack and distribute the PCR master mix in the 96-well plate according to the Taqman plate setup. Add the molecular grade water to the corresponding wells (i.e., in the negative controls) and transfer the plate with the chilled metal plate in the assembly laboratory. In the assembly laboratory, add the DNA samples starting with the extraction control to avoid any contamination at this stage. Close the tubes when the DNAs of a patient are pipette in order to avoid contamination by aerosols. Repeat this for all the patients' DNAs. When all the patients' DNAs are added, cover the corresponding wells and carefully add the positive control in the wells corresponding to the "inhibition control." It is important to avoid any contamination of the patients' test-tube with the positive controls which would lead to false-positive results. Exit the room with the plate in a chilled metal rack and transfer it to the amplification room where the thermocycler is located.
12. As described in Opota et al. 2015, if both PCR are positive (PCR1 and PCR2) in a respiratory sample, this indicates the presence of *C. psittaci* DNA [12]. Samples positive for PCR1 and negative for PCR2 can be considered positive for *C. psittaci* for respiratory specimens and positive for *C. abortus* for genital specimens.

Acknowledgments

We are grateful to all the technicians of the Diagnostic Microbiology Laboratory of the Institute of Microbiology of the Lausanne University Hospital for their technical contribution.

References

1. Lamoth F, Greub G (2010) Fastidious intracellular bacteria as causal agents of community-acquired pneumonia. Expert Rev Anti-Infect Ther 8(7):775–790. doi:10.1586/eri.10.52
2. Longbottom D, Coulter LJ (2003) Animal chlamydioses and zoonotic implications. J Comp Pathol 128(4):217–244
3. Yin L, Kalmar ID, Lagae S, Vandendriessche S, Vanderhaeghen W, Butaye P, Cox E, Vanrompay D (2013) Emerging *Chlamydia psittaci* infections in the chicken industry and pathology of *Chlamydia psittaci* genotype B and D strains in specific pathogen free chickens. Vet Microbiol 162(2–4):740–749. doi:10.1016/j.vetmic.2012.09.026
4. Magnino S, Haag-Wackernagel D, Geigenfeind I, Helmecke S, Dovc A, Prukner-Radovcic E, Residbegovic E, Ilieski V, Laroucau K, Donati M, Martinov S, Kaleta EF (2009) Chlamydial infections in feral pigeons in Europe: review of data and focus on public health implications. Vet Microbiol 135(1–2):54–67. doi:10.1016/j.vetmic.2008.09.045
5. Geigenfeind I, Vanrompay D, Haag-Wackernagel D (2012) Prevalence of Chlamydia psittaci in the feral pigeon population of Basel, Switzerland. J Med Microbiol 61(Pt 2):261–265. doi:10.1099/jmm.0.034025-0
6. Reinhold P, Ostermann C, Liebler-Tenorio E, Berndt A, Vogel A, Lambertz J, Rothe M, Ruttger A, Schubert E, Sachse K (2012) A bovine model of respiratory *Chlamydia psittaci* infection: challenge dose titration. PLoS One 7(1):e30125. doi:10.1371/journal.pone.0030125
7. Ostermann C, Ruttger A, Schubert E, Schrodl W, Sachse K, Reinhold P (2013) Infection, disease, and transmission dynamics in calves after experimental and natural challenge with a bovine *Chlamydia psittaci* isolate. PLoS One 8(5):e64066. doi:10.1371/journal.pone.0064066
8. Senn L, Greub G (2008) Local newspaper as a diagnostic aid for psittacosis: a case report. Clin Infect Dis 46(12):1931–1932. doi:10.1086/588562
9. Asner SA, Jaton K, Kyprianidou S, Nowak AM, Greub G (2014) *Chlamydia pneumoniae*: possible association with asthma in children. Clin Infect Dis 58(8):1198–1199. doi:10.1093/cid/ciu034
10. Greub G, Sahli R, Brouillet R, Jaton K (2016) Ten years of R&D and full automation in molecular diagnosis. Future Microbiol 11:403–425. doi:10.2217/fmb.15.152
11. Welti M, Jaton K, Altwegg M, Sahli R, Wenger A, Bille J (2003) Development of a multiplex real-time quantitative PCR assay to detect *Chlamydia pneumoniae*, *Legionella pneumophila* and *Mycoplasma pneumoniae* in respiratory tract secretions. Diagn Microbiol Infect Dis 45(2):85–95
12. Opota O, Jaton K, Branley J, Vanrompay D, Erard V, Borel N, Longbottom D, Greub G (2015) Improving the molecular diagnosis of *Chlamydia psittaci* and *Chlamydia abortus* infection with a species-specific duplex real-time PCR. J Med Microbiol (2015), 64, 1174–1185. doi:10.1099/jmm.0.000139
13. Jaton K, Brouillet R (2011) Full automation of nucleic acid extraction with the dual starlet (hamilton)-magnapure 96 (roche) system: a preliminary experience. ECCMID 2011, Milano

Chapter 12

Real-Time PCR to Identify Staphylococci and Assay for Virulence from Blood

Charles E. Okolie

Abstract

The genus *Staphylococcus* includes pathogenic and non-pathogenic facultative anaerobes. Due to the plethora of virulence factors encoded in its genome, the species *Staphylococcus aureus* is known to be the most pathogenic. *S. aureus* strains harboring genes encoding virulence and antibiotic resistance are of public health importance. In clinical samples, however, pathogenic *S. aureus* is often mixed with putatively less pathogenic coagulase-negative staphylococci (CoNS), both of which can harbor *mecA*, the genetic driver for staphylococcal methicillin-resistance. In this chapter, the detailed practical procedure for operating a real-time pentaplex PCR assay in blood cultures is described. The pentaplex real-time PCR assay simultaneously detects markers for the presence of bacteria (*16S rRNA*), coagulase-negative staphylococcus (*cns*), *S. aureus* (*spa*), Panton-Valentine leukocidin (*pvl*), and methicillin resistance (*mecA*).

Key words Staphylococcus, Virulence, Staphylococcal proteinA, Panton-valentine leukocidin, Blood culture, Differential diagnosis

1 Introduction

The genus *Staphylococcus* comprises facultative anaerobes, that is, they are capable of growth both aerobically and anaerobically. Staphylococci were historically delineated into two types; namely pathogenic and non-pathogenic. Both types of *Staphylococcus* spp. have been detected in healthy subjects in the community and in hospital patients [1, 2]. The staphylococci are relevant in public health because of their association with diverse infections including pneumonias, skin diseases, septicaemias, enteric infections, toxic shock syndromes, wound infections, bed sores, and urinary tract infections. The genomes of many staphylococci harbor genes encoding antibiotic resistance and horizontal genetic transfer (HGT) are well reported between different species; the *mecA* gene that codes for low-affinity penicillin-binding protein (PBP2a) is a typical example [3, 4]. *S. aureus* genomes are especially known for making a variable number of molecules associated with virulence

Kimberly A. Bishop-Lilly (ed.), *Diagnostic Bacteriology: Methods and Protocols*, Methods in Molecular Biology, vol. 1616, DOI 10.1007/978-1-4939-7037-7_12, © Springer Science+Business Media LLC 2017

and disease pathogenesis [5]. Some of the virulence proteins associated with *S. aureus* are encoded by species-specific genes such as *spa* (the gene that codes for staphylococcal proteinA, Spa), *coa* (the gene coding for coagulase, Coa), and *nuc* (the gene encoding staphylococcal thermonuclease, Nuc): these molecules have only been found in *S. aureus*, and not in the other staphylococci often called coagulase negative staphylococci (CoNS). Some *S. aureus* strains, however, harbor other virulence factors not commonly shared within the species; typical examples are the *tst* gene that codes for staphylococcal toxic shock toxin (Tst), the co-transcribed bi-component *lukSF-PV* (also called *pvl*) operon for the Panton-Valentine leukocidin toxin (PVL), and the enterotoxins that are very diverse. For this reason, it is important in diagnostic bacteriology to identify staphylococci and to assay for virulence, especially in blood. Although staphylococci have been identified by phenotypic methods over the years, which is still the current practice in some laboratories, the emergence of antibiotic resistance and virulence in many bacterial species including the staphylococci challenge the reliability of agar-based bacteriological tests [6, 7]. The way around this is that several laboratories have begun introducing diagnostic tests based on the polymerase chain reaction (PCR), including multiplexed assays that detect more than one DNA target in a PCR well [8, 9]. One of the off-shoots of conventional PCR is the application of florescent dyes to monitor amplification products as they accumulate. This is the basic principle of the modern real-time PCR that circumvents the need for resolution of PCR amplification products in gel electrophoresis. In this chapter, the steps toward the development and validation of a real-time PCR assay capable of detecting and quantifying five DNA targets for the identification of ubiquitous bacteria (*16S rRNA*), *S. aureus* (*spa*), the PVL operon (*pvl*), methicillin resistance (*mecA*), and marker for CoNS (*cns*) are described. By virtue of specificity (100%), sensitivity (Limit of detection, LoD = 1 colony forming unit per ml), simplicity, and accuracy, this method will be useful in clinical diagnostic settings.

2 Materials

2.1 Specialist Equipment

1. **DNA work station**: DNA work station is required for the preparation of DNA and solutions. L020—XL—GC (Kisker Biotech GmbH & Co. KG, Germany) is recommended (*see* **Note 1**).
2. **Centrifuge for real-time PCR multiwell plate (MWP)**: You will need to use a centrifuge to pull down contents after filling the MWP and before loading into the real-time PCR instrument for a run. Examples are the Beckman Coulter centrifuges

Table 1
Excitation-emission filter combinations of the Roche LightCycler480 for the fluorophores detected and their target DNA sequences (*see* Note 3), respectively (reproduced from author's previous work [11]

Target DNA (and amplicon size)	DNA probe Fluorophore	Channel on Machine	Wavelength (nm)	
			Excitation	Emission
16SrDNA (174 bp)	Cyan-500	Cyan-500	450	500
pvl (118 bp)	6FAM	FAM	483	533
spa (101 bp)	VIC	HEX	523	568
CoNS (204 bp)	LCR610	LCR610	558	610
mecA (155 bp)	LCR670	LCR670	615	670

(Allegra® X-30, Allegra® X-22, and Allegra® 21 series) with S2096 Swinging-Bucket Rotor and the FastGene® Plate Centrifuge (Nippon Genetics Co. Ltd., Japan). Please stick to whatever brand name you have in your laboratory. This item of equipment is necessary as the optical system of the machine acquires fluorescence best without air in the fluid.

3. **Real-time PCR system** (*see* **Note 2**): The assay described in this chapter was developed using the LightCycler® 480 (abbreviated: LC480) from Roche (Roche Applied Science, Penzberg, Germany) [10]. However, if your laboratory has another real-time PCR instrument, it does not matter. Table 1 shows excitation–emission filter combinations of the LightCycler® 480 instrument I filter set recommended for fluorophores that can also be used with various different real-time PCR detection formats (*see* **Note 3**).

2.2 Bacterial Strains and Bacteriological Media

Bacteriological media and consumables are necessary for confirming the real-time PCR. In the UK, Oxoid (Oxoid, Basingstoke, UK; recently acquired by ThermoFisher) is the major supplier of dehydrated media. However, the experimenter is advised to adhere to locally available media and consumables as the staphylococci are not very difficult to grow in basal media.

1. Reference-type culture staphylococcal strains: Be sure to have reference type cultures of staphylococci and other bacteria necessary for assay validation. Based on the genetic targets of the assay (bacterial *16S rRNA*, *spa*, *cns*, *pvl*, and *mecA*), the least would be a coagulase negative staphylococcus such as *S. epdermidis* and a *S. aureus* strain harboring *pvl* operon. It does not matter if the *mecA* is in a CoNS or *S. aureus* genome. The Network on Antimicrobial Resistance in *S. aureus* (NARSA) is

particularly good because information on the gene loci essential for this assay is provided on their website.

2. Brain heart infusion (BHI) broth.
3. BHI agar.

2.3 Real-Time PCR

1. **Real-time PCR mastermix**: Mastermix is often supplied by the manufacturer of the real-time instrument. The pentaplex real-time PCR assay described in this chapter was developed using LC probe mastermix supplied alongside the LC480 machine.
2. **Triton X-100 lysis buffer**: (100 mM NaCl, 10 mM Tris–HCl [pH 8], 1 mM EDTA [pH 9], and 1% Triton X-100). All reagents are from Sigma Chemicals, St. Louis, MO, USA.
3. **PCR grade water**: We used the PCR water supplied with the LC480 instrument. You can always buy from Sigma (Sigma-Aldrich, USA) or other suppliers.
4. **Oligonucleotide primers and probes**: The sequences of the primers and probes are provided in Table 2. The source of oligonucleotide primers and probes is important. Probes synthesized by TIB Molbiol (TIB Molbiol, Berlin, Germany) and HPLC purified primers from Sigma (Sigma-Aldrich, UK) were used in the assay reported in this chapter [11, 12]. Probes and primers may be received lyophilized or in solution—they all work very well.

3 Methods

3.1 Handling and Preparation of Oligonucleotide Primers and Probes

1. From the supplied oligonucleotide primers and probes, calculate the required dilution steps to arrive at the predetermined concentrations (Table 3). However, you may need to optimize on your machine.
2. Although primers and probes withstand freezing and thawing, it is advisable to prepare them in aliquots of low concentration, to reduce repeated freezing-and-thawing cycles.
3. Avoid exposing oligonucleotides to direct sources of sunlight and heat other than the PCR cycling process.
4. Follow the manufacturer's instruction. Whenever in doubt, ask.

3.2 Handling and Preparation of PCR Mastermix and Solutions

1. Thaw the mastermix and other solutions including the template DNA suspension (if coming from the freezer) at room temperature. Please allow complete thawing. This way you are sure you are using a homogenous mix.
2. Mix gently by tilting. Vortexing hard is not recommended.

Table 2
Primer-probe sets required for the real-time pentaplex PCR assay and their target amplification sequences, respectively (reproduced from [12] with permission from Elsevier)

Target sequence	Amplicon size (bp)	Primer and probe identity[a]	Primer and probe[b] sequence 5′ → 3′
CoNS	204	cns-F	TATCCACGAAACTTCTAAAACAACTGTTACT
		cns-P	[c]LCR610-TATTAGACTACGCTGAAGCTGGTGACAACAT-BBQ
		cns-R	TCTTTAGATAATACGTATACTTCAGCTTTGAATTT
16S	174	16S–F	CTAGTAATCGCGGATCAGCAT
		16S–P	Cyan500-TAGCCGTCGAAGGTGGGACAAATGAT-BBQ[d]
		16S–R	GATACGGCTACCTTGTTACGACTT
mecA	155	mecA-F	TGGTATGTGGAAGTTAGATTGGGAT
		mecA-P	[e]LCR670-TTCCAGGAATGCAGAAAGACCAAAGCA-BBQ
		mecA-R	CTAATCTCATATGTGTTCCTGTATTGGC
pvl	118	lukSF-PV-F	TTACACAGTTAAATATGAAGTGAACTGGA
		lukSF-PV-P	FAM-AAGTGAAAGGACATAATTG-MGBNFQ[f]
		lukSF-PV-R	AGCAAAAGCAATGCAATTGATG
spa	101	spa-F	CAGCAAACCATGCAGATGCTA
		spa-P	HEX-TCAAGCATTACCAGAAAC-MGBNFQ
		spa-F	CGCTAATGATAATCCACCAAATACA

[a]F = forward primer; P = probe; R = reverse primer
[b]Probe sequence comprises of fluorophore-nucleotides-quencher
[c]5′-End of the probe was labeled LightCycler®Red 610 (LCR610)
[d]3′-End of the probe was labeled with BlackBerry Quencher (BBQ)
[e]5′-End of the probe was labeled LightCycler®Red 670 (LCR670)
[f]3′-End of the probe was labeled with a minor groove binder molecule and non-fluorescent quencher (MGB-NFQ)

Table 3
Constituent factors of the optimized PCR (reproduced from [12] with permission from Elsevier)

Component	Quantity
Roche LC probe mastermix	25.0 μl
Template DNA	10.0 μl
spa and pvl primers	0.75 μM
Bacterial 16SrRNA CoNS, and mecA primers	1.0 μM
spa probe	0.04 μM
pvl probe	0.08 μM
mecA, CoNS and 16SrRNA probes	0.2 μM
Roche PCR grade water	Make up to 50.0 μl PCR
Number of cycles	40

3.3 Incubation of Blood Cultures (See Note 4)

1. Incubate until blood culture bottles show positive signal or the presence of bacteria by Gram stain.
2. Process blood cultures for PCR within 12 h from positive signal.

3.4 Culture of Reference and Control Bacterial Strains

1. Resuscitate reference and control bacterial strains, presumably obtained from frozen (−80 °C) by inoculating into BHI broth. Incubate at 37 °C for 8–12 h.
2. From the BHI broth culture, streak out for discrete colonies on a BHI plate. Incubate overnight at 37 °C.
3. Use discrete colonies for simple phenotypic characterization including Gram staining, catalase, and coagulase tests.
4. The phenotypically characterized colonies from reference and control bacterial strains will be useful for inoculating real-time PCR wells.

3.5 Preparation of Bacterial DNA (Method 1: From Reference and Control Bacterial Strains in BHI Broth)

1. Prepare bacterial DNA using the method of Johnsson et al. [13].
2. From the BHI broth, transfer 0.5 ml of bacterial suspension into a 1.5 ml Eppendorf tube and heat for 10 min at 98 °C.
3. Following a quick centrifugation (13,000 × *g* for 20 s), transfer the supernatant into a fresh 0.5 ml Eppendorf tube for use as a source of template DNA for PCR assay.

3.6 Preparation of Bacterial DNA (Method 2: From Blood Culture Bottles)

1. Prepare bacterial DNA from blood culture bottles using the method of Louie et al. [14].
2. From positive signalled blood culture bottle, transfer an aliquot (100 μl) to 1000 μl of PCR grade water in a 1.5 ml Eppendorf tube. Mix by inversion, 4–5 times. Leave to stand on the rack in the DNA work station for 5–10 min.
3. Centrifuge at 16,000 × *g* for 1 min. Discard the supernatant.
4. Resuspend the pellet in 100 μl of Triton X-100 lysis buffer.
5. Add 5.0 μl of lysostaphin containing 1 mg of lysostaphin per ml (Sigma, USA). Mix and incubate at 37 °C for 10 min.
6. Boil this suspension for 10 min. Cool to room temperature for 5 min.
7. Centrifuge at 16,000 × *g* for 1 min.
8. Transfer the supernatant into a fresh 0.5 ml Eppendorf tube for use as a source of template DNA for PCR assay.

3.7 Biological Controls (See Note 5)

To institute biological controls for your in-house use, prepare the DNA from previously tested and confirmed bacterial strains including staphylococcal and non-staphylococcal, following the steps in Subheading 3.5. Load the control DNA in appropriately designated wells to run in your experiments. They should generate the expected amplification signals.

3.8 Preparing and Loading the 96-Well Plate

1. Seal the plate properly with a sealing foil by pressing it firmly to the plate surface using your hand or a scraper (e.g., the Sealing Foil Applicator provided with the instrument). Be careful with the sealing foil applicator to avoid scratching and creasing it on any object as this could cause optical interference.
2. Place the 96-well plate in a standard swing-bucket centrifuge, containing a rotor for multiwell plates with suitable adaptors. Balance it with a suitable counterweight (example: another MWP with water droplets in the wells—no need to ensure weight-to-weight accuracy as many modern centrifuges are optimized to handle such minor balancing differences). Centrifuge the plate at 1500 × *g* for 2 min. Check the wells for bubbles, and repeat if necessary.
3. To load the prepared MWP into the LC480 instrument, press the push button on the front of the instrument (located next to the instrument status LEDs): The MWP loader extends out of the right side of the instrument.
4. Place the MWP into the loading frame of the loader with the flat edge pointing toward the instrument. (The short plate edge with beveled corners points away from the instrument.)
5. Press the plate loading push button again to retract the loader with the inserted MWP into the instrument. You are now ready to start the run.

3.9 Operating the LightCycler® 480 Real-Time PCR Instrument

1. Using the mains switch located on the left side of the power box at the back of the instrument, turn the LightCycler® 480 instrument on. Two status LED lamps (*see* **Note 6**) are located at the front of the machine (left and right). Watch out for the following functional indicators of the LED lamps during instrument operation. When the LC480 instrument is initializing, both the left and the right LEDs are flashing. This lasts about 2 min. When the left is green and the right is orange, the instrument is turned on and ready. Instrument status is ready. No plate is loaded. The instrument automatically brings out the plate reel to receive the plate from you.
2. Place the 96-well plate on the reel. While the plate is loading, the left LED is green and steady while the right LED is orange and flashing. Next, the LEDs are green and steady indicating that the instrument is turned on, instrument status is ready, and the plate is loaded. And then both LEDs are green and flashing indicating that the instrument is running (*see Note*: The instrument is running does not mean the specific experiment you are after is running. It serves to show that the instrument is up and running and in a good condition to perform your experiment).

3. Turn on the control computer unit that could be a desktop or a laptop.
4. Log on to Windows XP.
5. You need to understand the LC480 software before operating further (*see* **Note 7**). To start the LC480 software, double-click the LC480 software icon on the computer.
6. In the *Login* dialog box, type your user name and password. The initial password for the admin user is LightCycler480.
7. Click the *Proceed* sign to proceed with the login. The application now displays the LightCycler® 480 Software *Overview* window (*see* **Note 8**) containing the *Status* bar at the top, the *Experiment Creation* and *Tasks* area on the right, and the *Global action* bar on the extreme right. From the *Overview* window, you can do the following:
   (a) Create a new experiment.
   (b) Create a new experiment from a macro or a template (Fig. 1).
   (c) Open an existing object.
   (d) Switch to other software modules such as the *Navigator* or the *Tools* section.
8. In the "Log on to" field, the last database selected on the computer is displayed by default (Example: In our laboratory, the experiment reported in this chapter is saved as "Staphylococci—Charles"). If several databases are available and you want to log on to a different database than the one displayed in the "Log on to" field, select the Options button. The Login dialog box expands to show a selection list of all available databases. Select a database from the list. The overview window has several areas loaded with different tasks and functions.
9. Click the *Proceed* sign to proceed with the login. The application now displays the LightCycler® 480 Software *Overview* window containing the *Status* bar at the top, the *Experiment Creation* and *Tasks* area on the right, and the *Global action* bar on the extreme right.

   Understand the following areas of the *Overview* window:

   (a) *Status* bar
   This area displays information about the currently active object and allows the user to select an object to view from a list of currently open objects. The fields in the *Status bar* and their functions are described below:

   - The User Field shows the name of the registered user currently logged on to the machine. This is made possible because every user registered on the machine has to sign in before performing any operation.

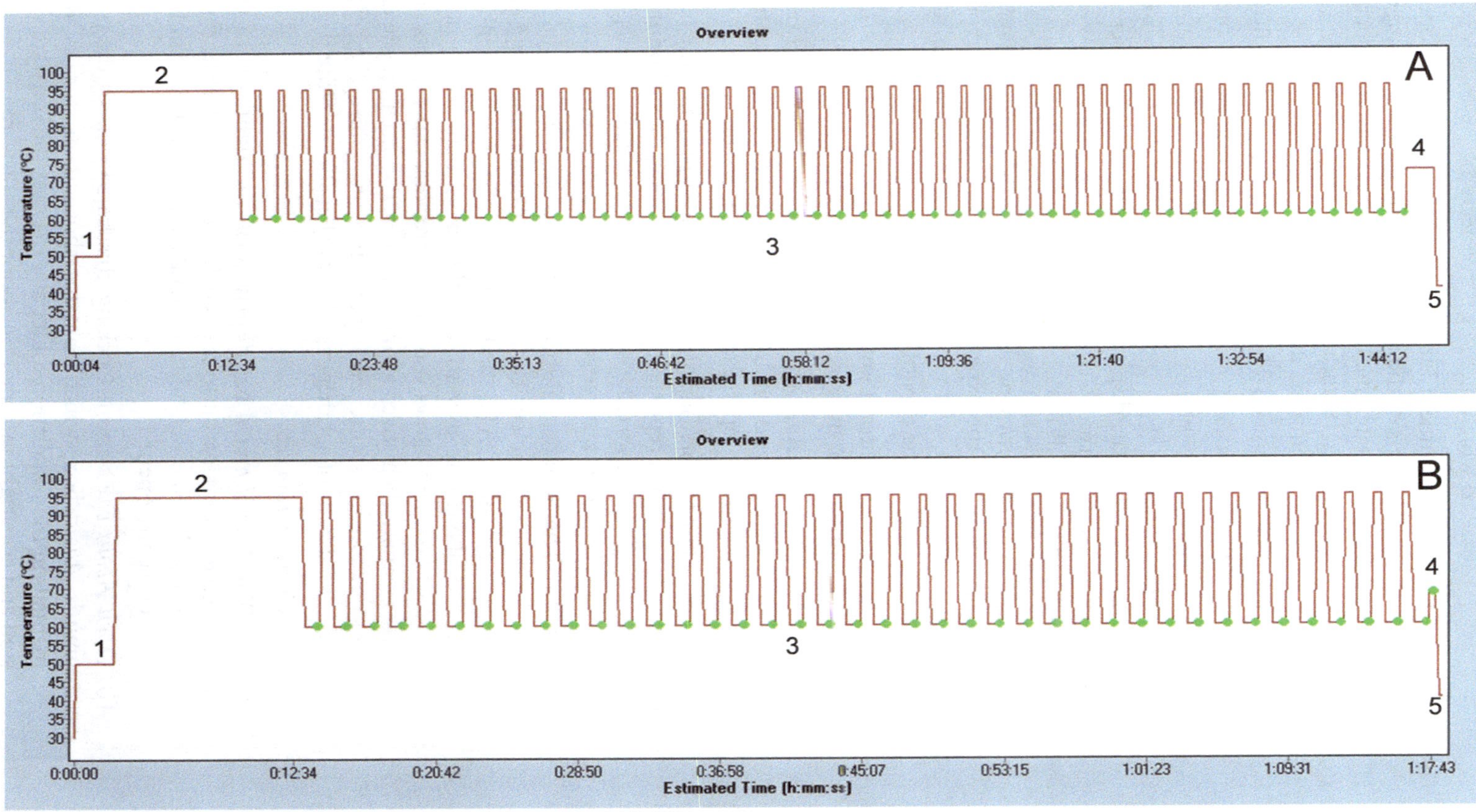

Fig. 1 Obtaining a shorter cycling time by editing the cycling conditions of another experiment. (**a**) Overview of a 50-cycle assay that finishes in 1 h 44 min 12 s. (**b**) Overview of another assay with fewer (40) cycles obtained by editing (**a**) and it finishes in 1 h 17 min 43 s, a gain of approximately 27 min (reproduced from author's previous work [11])

- The Database Field shows the name and type of database to which the user is connected.
- The Instrument Field shows the name and the operational status of the instrument the user is working on. Possible states include: Not Connected, Connected, Initializing, Standby (MWP loaded), Standby (No MWP), Running, and Error.
- The Windows Field shows a pull-down menu listing all currently open windows. The pull-down menu enables the user to switch between windows.

(b) *Module* bar

The *Module* bar, displayed on the left side of the screen, has six permanent buttons and their functions are described as follows:

- The Experiment button opens the Run module including the experiment protocol, charts of experiment data, and notes entered by the logged in user running the experiment.
- Subset Editor allows the user to group samples into subsets for analysis and for reports.
- Analysis Button launches the *Analyses Overview Window* and enables the user to either create new analysis or open existing analyses.
- The Report Button allows the user to define the contents of a report. It also enables viewing and printing of the report. However, the rule is that the Report Button is not active until an experiment is saved.
- The Summary Button opens the summary module of the experiment containing information including the NAME, DATE, and OWNER of the experiment. It allows the user to save the experiment as a macro. A macro allows you to edit previous experiments and have them saved with new names. For instance, you can obtain a shorter cycling time by editing the cycling conditions of another experiment (Fig. 1) (*see* **Note 9**).

(c) *Global action* bar

The *Global action* bar displayed on the right side of the screen contains buttons used for general software functions. Their availability depends on the active window currently opened.

10. *Open About box* button. Clicking this button opens the program's *About* box, which contains shortcuts to the LC480 manuals in the installation folder and displays the software version and copyright information about the software.

11. Use the "Selection and Navigation Features" (*see* **Note 10**).
12. To create and execute a query, select the *Query* tab in the *Navigator* window. In the *Object Type* box, select the type of object to be retrieved. (Optional) Enter the name of the item to be retrieved or the owner of the item, if known. *You may throw the wildcard "*" to search for any character string.*
13. To create or modify an object (in the navigator), select *Modification Date* or *Creation Date* to specify which date you want to use in the query. The Modification Date and the Creation Date choices are mutually exclusive (i.e., you can search for one or the other, but not both). Select a date range for the search. You can specify the number of months or days before the current date or you can select a beginning and ending date in the past.
14. For any possible object type, you can also select a target folder from the *Folders* tab. Check the *Scan Sub-folders* box to include all subfolders within a directory in the search.
15. In using the navigation features, understand that the Roche folder contains some useful standard objects:
    - Demo experiments in the Experiments subfolder (*see* **Note 11**).
    - Demo run templates in the Templates subfolder including predefined subsets in the Subsets subfolder. The run templates folder contains templates for use with the LC480 instrument I or LC480 instrument II.
    - Color Compensation objects including the universal Color Compensation objects, and a demo melt object for Endpoint Genotyping analysis in the Special Data folder.
    - To modify a Roche object, you must first create a copy by exporting and importing it to your own user folder.
    - Administration folder that contains objects for user groups, user roles, user accounts, and security policies.
16. Click the *Search* button. Results are displayed in the pane to the right of the search criteria. The results include the following:
    - Object name.
    - Object type.
    - Date the object was created.
    - Date the object was last modified.
17. You can sort the result list in ascending or descending order by clicking the corresponding column head. If you select an object in the list, the full path to the object is displayed in the *Status* bar at the bottom of the Results pane. If the selected object is an experiment or a macro, a summary of object information is displayed in the Object Summary pane (*see* **Note 12**).

18. To open an object, double-click the object name.
19. Select and Deselect Samples. The Sample Selector is self-explanatory (*see* **Note 13**).
20. To set the selection status of enabled samples in the MWP image, click a sample to select it. Press the <Ctrl> key and click a selected sample to deselect it. If the MWP image and the *Subset* combo box are used together, selecting a subset enables only the samples in the subset and automatically selects them.
21. Click and drag on a deselected well to select all wells in the drag region. Press the <Ctrl> key and click and drag on a selected well to deselect all wells in the drag region.
22. Click row or column headers to select the corresponding rows or columns. Press the <Ctrl> key and click row or column headers to deselect the corresponding rows or columns. The display of the sample table corresponds to the selection of rows or columns in the sample selector.
23. Click the square in the upper left corner of the MWP image to toggle the selection status of the entire plate. If a legend is included in the *Sample Selector* you can use the *Legend Property Selector* to select which legend options are displayed. The options provided in the *Legend Property Selector* combo box depend on the context.
24. Click a colored icon in the legend to toggle the selection status of the corresponding wells. Selection of the legend icons is synchronized with the selection in the MWP image. A legend icon appears selected if all members of the group are selected in the MWP image. It will not appear selected if any group member is not checked.
25. Double-click a legend icon to select all items in the corresponding group and deselect all items not in the group.
26. To add wells to or remove them from a selection, press the <Ctrl> key and click a well, row, or column or click and drag a rectangle.
27. Scroll the MWP Image.

    The MWP image contains horizontal and vertical scroll bars to allow you to scroll and *see* any part of the image. When you are scrolling, the column and row headers remain fixed.
28. Zoom in and out of the MWP Image.

    You have several possibilities for zooming in on or out of the MWP image:

    - With the zoom buttons.
    - With the slider bars.
    - Manually setting the zoom by dragging the margins of a column or row.

The MWP image provides two slider bars for zooming in a horizontal or vertical direction:

- With the horizontal slider at the rightmost position (minimum zoom) the MWP image displays a single column.
- With the horizontal slider at the leftmost position (maximum zoom) the MWP image displays all columns.
- With the vertical slider at the topmost position (maximum zoom) the MWP image displays a single row.
- With the vertical slider at the bottommost position (minimum zoom) the MWP image displays all rows.

As well as zooming with the buttons or sliders, you can manually set the horizontal or vertical zoom by dragging the right margin of a column or the bottom margin of a row.

29. **(Optional) Print the MWP Image**
The MWP image provides a *Print* button that allows you to print the visible section of the image. The print is scaled to a single page.

30. Use the *Sample Table* and *Sample Selector* to Select or Deselect samples for display in an analysis chart or to include/exclude a sample from analysis (*see* **Note 14**). You can select one or all of the samples in the *Sample Table* for display in an analysis chart, but you cannot change any of the information displayed. Selected samples are highlighted. To add or remove samples from the selection in the *Sample Table*, use the standard windows shift-click and ctrl-click features.

31. **Include or Exclude Samples** (*see* **Note 15**). Further, samples can be included into or excluded from analysis. To include a sample, mark the *Include* box in the left table column. Status of the *Include* box is changed by double-clicking or by pressing the <Space> key. Using the *Include* option, you can, for instance, decide which standards are used to calculate the standard curve in Absolute Quantification analysis.

32. Export and Import Objects and Experiment Raw Data. To store objects outside the LC 480 Software database or to transfer objects between databases, you must export the LC 480 Software files (*see* **Note 16**).

33. For experiments you can also export the raw data in a tab-delimited text format (Experiment Text File (*see* **Note 17**)).

34. To start the experiment run, click the *Start Run* button on the Run Protocol or the Data tab. You can only start an experiment run when an instrument is connected. The Start Run button is only available if a MWP has been loaded. You are prompted to save the experiment. Enter an experiment name and browse to a folder where you want to save the experiment. If you close the dialog without saving the experiment, the

run will not start. A status bar on the *Data* tab indicates the progress of the running experiment (*see* **Note 18**).

35. To view data for specific samples, select one or more samples in the *Sample Editor* or *Sample Table.* If needed, you can modify chart settings during the experiment run.
36. (Optional) To adjust or stop the program during the run, click End Program to stop the current program and skip to the next program in the experiment protocol. Performing this task ensures that the data are complete and an analysis can be performed.
37. To stop the run, click the *Abort Run* button. (The *Abort Run* button replaces the *Start Run* button during the run.) Performing this task results in incomplete data, no analysis can be performed.
38. When the experiment is finished, a status message displays "*Run complete.*"
39. When the run is finished, open the plate loader again to remove the MWP. Hazard (*see* **Note 19**).
40. The next stage is the data analysis (*see* **Note 20**). Click *Sample Editor* in the *Module* bar to open the *Sample Editor*, and complete sample information, if necessary (*see* **Note 21**).
41. To perform an analysis, open the experiment you want to analyze in the LightCycler® 480 Software main window.
42. In the *Module* bar, click *Sample Editor*. If you have not already entered sample information, enter information to identify the samples.
43. Enter the analysis-specific sample information. The kind of information you can enter depends on the type of analysis.
44. Click *Analysis* on the LightCycler® 480 Software *Module* bar. The *Analysis Overview* window opens. The *Analysis Overview* window displays the *Create New Analysis* list and *Open Existing Analysis* list (if an analysis was created before).
45. Select the analysis type from the *Create New Analysis* list. The *Create new analysis* dialog opens. Here, you can again define the analysis type and select an analysis subset. If your experimental protocol should contain several programs that are suited for the selected analysis type, select one from the Program list. If you wish, you can change the analysis name (the default name is "analysis type for subset name"). Click the *Proceed* button (*see* **Note 22**).
46. In the *Analysis* window enter or adjust the parameters. Use the *Color Compensation* multi-select button to turn Color Compensation on or off and to select a Color Compensation object. Use the Filter Comb button to select the fluorescence channel you want to analyze.

47. Click Apply Template to display the Apply Template dialog box. Select a Template from the list and click the "proceed" button. The template settings are applied to the new experiment protocol.
48. Select the options specific to the analysis type. If you change information in the Sample Editor (except sample name, sample note, or target name) after performing the analysis, you must recalculate the analysis results using the updated values from the Sample Editor. In this case, the Calculate button becomes active again.
49. You can add more than one analysis to an experiment, including multiple instances of the same analysis type. Click the "add" in the *Analysis* toolbar.
50. Click the "save" button in the *Global action* bar to save the analysis results as part of the experiment.
51. To select samples to include or exclude in result calculations, select the checkbox next to a sample name to generate analysis results for the sample. By default, all samples are checked at the beginning of an analysis. Double-click a sample checkbox to deselect or reselect it. To check or uncheck a group of samples simultaneously, highlight the range of samples, and press the <Space> bar. This toggles the check marks on or off in all the selected sample boxes.
52. To select samples to view in charts, note the color of the sample in the sample list (samples are color-coded), and look for the color on the chart. Alternatively, place the mouse pointer over a line on a chart to display a small box containing the name of the sample represented by the line. When you highlight a sample name in the sample list, data from the selected sample is displayed in the analysis charts. By default, all samples are selected when you first open the analysis window.
53. You can redraw the analysis graphs using selected samples:
    - To select one sample, highlight the sample name in the sample list.
    - To select multiple samples, press the <Ctrl> key while clicking the sample names.
    - To select a contiguous set of samples, click the first sample name, and press the <Shift> key while clicking the last sample name in the set.
    - To select all samples, press <Ctrl-A>.
54. You can remove or rename analyses saved with your experiment if your user account has the Expert User or Local Administrator role. You may also be able to remove or rename analyses stored with experiments of other users, depending

on the access privileges associated with your user account. To remove an analysis from an experiment:

- Select an analysis from the *Analyses* bar.
- Click the *Remove Item* button in the *Analysis* toolbar.
- You are prompted to confirm your choice.
- Click the *Remove Item* button in the *Analysis* toolbar to remove the analysis.
- Click the *Save Item* button to save the experiment without the analysis.

55. You can rename the analysis associated with an experiment (renaming is helpful if you have more than one analysis of the same type associated with the experiment). Select an analysis from the *Analyses* bar. Click the *Rename Item* button in the Analysis toolbar. The *Edit Name* dialog opens. Enter a new name and click the *Proceed* button. (Remember: Names must be unique in each database folder).
56. Export analysis results (*see* **Note 23**). Click the *Result Batch Export* Navigator control button. The *Batch Export* wizard opens. On the *Select Folders* tab of the wizard, select a source folder in the *Available Folders* list and add it to the *Selected Folders* list. Check the *Scan Sub-folders* option to include all subfolders within the source directory.
57. On the *Target* tab, select the destination directory and the name of the output file. Click the *Browse* button to browse for a destination. Alternatively, type the path of the destination directly into the input field. If the specified output file already exists, the wizard will ask you to confirm before overwriting the existing file.
58. On the *Analysis Type* tab, select the type of analysis to be exported.
59. On the *Start* tab, you can review your settings and start the export process. The Reset button on the Start tab is active only after an export is complete. Clicking the Reset button resets the results of the previous export, so the export can be repeated.
60. On the *Status* tab, you can view the status of the export process. While the export is in progress, the *Stop* button is active. You can abort the export by clicking the *Stop* button.
61. The *Done* tab displays a summary of the batch export. Click the *Done* button to close the wizard. One of the key features of the LC480 system is that the software is designed to display detection signal(s) from one selected signal emission channel at any one time (Fig. 2). The LC480 system does not allow for simultaneous display of detection signals from all the emission channels (*see* **Note 24**).
62. The analyzed data is now ready for presentation, publishing, or other uses.

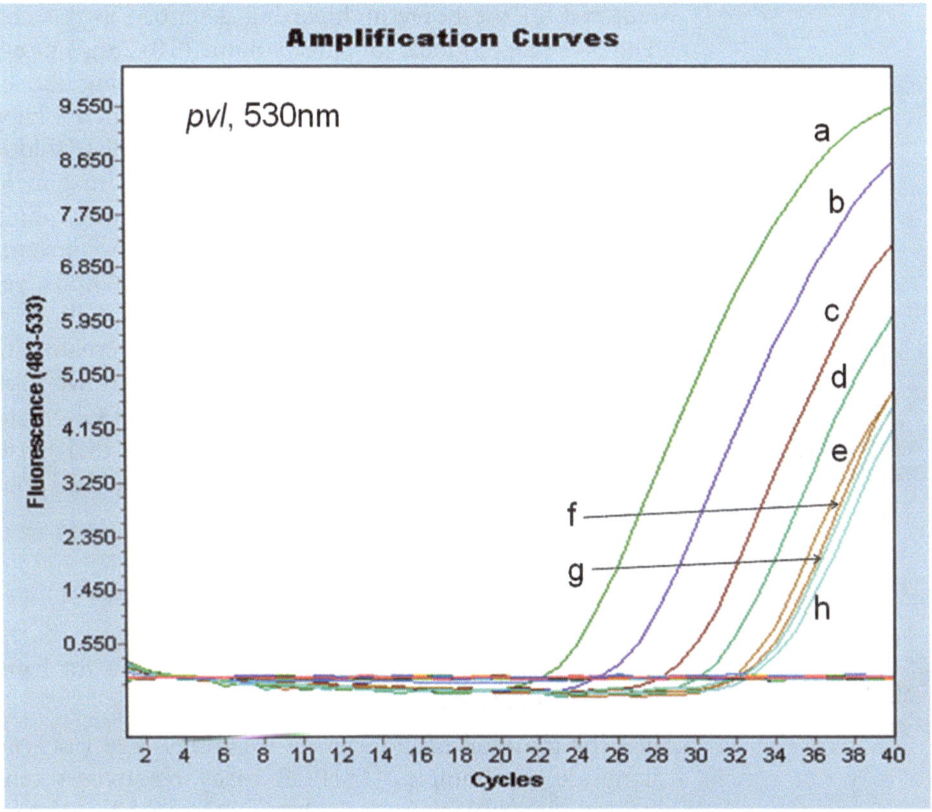

Fig. 2 Amplification curves for pvl detected via the FAM (530 nm) emission channel showing the fluorescence versus Cp for measured bacterial DNA load. 1.0 × 10^7 CFU [Cp = 21.07] (**a**), 1.0 × 10^6 CFU [Cp = 23.15] (**b**),1.0 × 10^5 CFU [Cp = 26.28] (**c**), 1.0 × 10^4 CFU [Cp = 29.25] (**d**), 1.0 × 10^3 CFU [Cp = 32.90] (**e**),1.0 × 10^2 CFU [Cp = 33.03] (**f**),1.0 × 10^1 CFU [Cp = 33.32] (**g**), 1.0 CFU [Cp = 33.82] (**h**), respectively (reproduced from author's previous work [11]

4 Notes

1. If your laboratory has a biological safety cabinet (BSC) or microbiological safety cabinet (MSC) that has **UV or Air-cleaner**, that would suffice.
2. The LC480 is described briefly. The LC480 instrument is available in two versions, one with a thermal block cycler for the LightCycler® 480 MWP 96-wells, the other for the LightCycler® 480 MWP 384-wells. Additionally, both LightCycler® 480 Instrument versions are available with two different thermal block cyclers: LightCycler® 480 Thermal Block Cycler Unit Silver and LightCycler® 480 Thermal Block Cycler Unit Aluminum. Both thermal block cyclers can be used with LightCycler® 480 Instrument I and II. Although you can purchase each version of the thermal block cycler as an exchangeable accessory (LightCycler® 480 Block Kit 96 or 384), we recommend the 96 block for the large volume of reaction (40–50 μl)

required for the heavy multiplexing described in this chapter. The 384 block is ideal for small volume (10–25 μl) monoplex reactions. We do not recommend interchanging the blocks because it often refuses releasing the holder. When it happens, calling the engineers is often not the best immediate solution as they would take time to get to the laboratory. Imagine losing the time, money, and materials invested in preparing a multiwall plate full of control and test samples. To circumvent this challenge, Roche later developed the 05015278001 version (96-well) and the 05015243001 version (384-well). Furthermore, following the launching of the LC480, alternative instruments capable of detecting DNA target through five floescent emission channels have joined the real-time PCR market—examples are the Rotorgene 6000 real-time PCR machine (Corbette, Australia) and the QuantStudio™ 6 and 7 Flex Real-Time PCR series (ThermoScientific, USA).

3. Besides the dyes listed in Table 1, all dyes that are compatible with the excitation and emission filter wavelengths can be measured by the LightCycler® 480 Instrument I.

4. It is important to follow the safety instructions for handling biological fluids and tissues.

5. It is the tradition of life sciences experiments to run controls alongside test samples. In PCR many researchers tend to include "internal controls." One commercial real-time PCR assay for PVL-MRSA that we operated clinical trials for (name withheld) was rubbished by the incorporation of an amplifiable internal control and oligonucleotides with which the internal control yielded amplification signal. However, we found that whenever CoNS was assayed, all the channels in the assay failed to detect the anticipated product (signal). In this pentaplex real-time PCR assay, the oligonucleotides targeting the bacterial *16S rRNA* gene serve the purpose of detecting this gene as a positive control for all wells containing bacterial DNA and it never failed in our hands. Diagnostic assay users are interested in assays that work not assays loaded with interfering substances called internal controls.

6. The LEDs are first signposts of instrument wellness or malfunction.

7. The instrument manual and the commands are loaded in the computer which is driven by the LC480 software. To continue with the operation, understanding the LightCycler® 480 software and the user interface is necessary—reason: when the user understands the software that drives the instrument, then it is easy to perform successful experiments with the machine. LightCycler® 480 Software controls the LightCycler® 480 instrument using information you provide in an experiment

protocol. Raw data gathered by the software during real-time online PCR monitoring can then be analyzed utilizing the various analysis software modules and their functionalities. In this configuration, all software components are preinstalled on the LightCycler® 480 control unit connected to the LightCycler® 480 instrument. Each configuration (instrument and connected control unit) functions as an independent system with its own databases and its own set of user accounts. The location and destination folder of the database engine and database files is usually predefined for the installation process. All data gathered by the LightCycler® 480 System are stored in the database to guarantee security for the data and data integrity. No manipulation of stored data and no access to raw data are possible. Analysis and editing of data can only be done within the LightCycler® 480 software. The user interface of the LightCycler® 480 Software displays some common elements including buttons with defined functionality which you will find on nearly all software screens. Furthermore, general button design conventions imply the function behind each button by using specific button indicators. Placing the mouse pointer over an icon or button displays a short description of its function and its keyboard shortcut (if available).

8. From the *Overview* window, you can create a new experiment, create a new experiment from a macro or a template, open an existing object, or switch to other software modules such as the *Navigator* or the *Tools* section. Understand the following areas of the *Overview* window:

   (a) Editor frame
   The Editor Frame is the central area where the modules are displayed. The frame may contain several sections that can be resized individually. You can resize a section by dragging the splitter bar on the border between two sections to hide or show the section. The arrows on the splitter bar indicate which area of the *Editor* frame will be affected. Clicking a splitter bar will hide the corresponding area.

   (b) Message area
   The *Message* area displays status messages, errors and warnings. The *Message* area consists of the following parts: Alarm **icon on the left**: The color of this icon changes depending on the severity of the alert:

   - Gray = normal information.
   - Yellow = warning.
   - Red = alarm condition.

   (c) Understand the "*Text field in the middle.*" The text field in the middle displays messages, including the type, date, and time of message and the message text. Right-click a message entry to open the corresponding context menu. Select *Show log* in the context menu to display the log file and open the *Error Log* tool. Select *Clear selected Messages* to delete the selected messages from the *Message* area. Select *Clear Messages* to delete all messages from the *Message* area. Double-click a message entry or select *Details* in the context menu to display detailed information.

9. The temperature regime was a continuous flow that could be shown to consist of the five distinct steps outlined below:

   - Step 1 was a single cycle of 2 min at 50 °C for UNG.
   - Step 2 comprised a single cycle of 10 min at 95 °C (initial template denaturation and activation of polymerase).
   - Step 3 comprised 40 cycles of two-temperature cycling consisting of 15 s at 95 °C for initial template denaturation and 40 s at 60 °C for amplification (annealing and polymerization) with acquisition of fluorescence signal at 60 °C.
   - Step 4 was a single stage of 5 s at 68 °C for maximum signal acquisition from all the channels, as observed from the results of the CC object.
   - At step 5, the system cooled down to 40 °C for 5 s for the plate to be removed while the data was ready for analysis.

10. This section describes the object selection, navigation, and query elements of the LightCycler® 480 Software:

    - Navigator
      The Navigator is similar, but not identical to the Windows Explorer of your computer. The Navigator displays data that are stored in a database not in the Windows file system. The *Navigator* window provides access to items stored in the LightCycler® 480 database. Items include experiments, user accounts, instrument, macros, etc. The Navigator allows you to open experiments and related items (such as preferences, macros, and special data) as discrete objects. All items in the Navigator are organized in folders in a tree-like structure (similar to Windows Explorer) and are sorted alphabetically within their folders. You can expand and contract folder views and highlight the object you want to select. In addition, you can use the *Query* tab to search for specific LightCycler® 480 Software objects in the database by entering search parameters.

      The *Navigator* window is structured into four areas:

      1. Tree pane.
      2. Object summary pane.
      3. Navigator controls.
      4. Query tab.

Tree Pane

The Navigator *Tree* pane displays a hierarchical tree view of the objects stored in the currently active database. The top object in the tree is always "Root." The tree is used in a similar manner as for Windows Explorer. The Navigator *Tree* pane always includes the following default folders and objects:

- User folders (including the *System Admin* folder and folders for each user account).
- Each user folder contains default subfolders, such as a folder for experiments.
- *Roche* folder that contains experiments, templates, and macros from Roche that can be used by anyone with access to LightCycler® 480 Software.
- To show or hide items under a folder, double-click the folder name or click the plus (+) or minus (−) sign next to the folder. Right-clicking an object in the *Tree* pane opens a context menu with the actions currently available for the object. For more information on the actions, *see* section Navigator Controls.

Object Summary Pane

The Navigator *Object Summary* pane displays experiment summary data if the currently selected object is an experiment or a macro.

Navigator Controls

In combination with the Tree pane, the Navigator control buttons allow you to work with objects in the database and to import and export objects. The buttons of the navigator Controls and their functions are described. The availability of the Navigator control buttons depends on your user role and on the database you have logged onto. A research database allows experiments and experiment-related objects to be renamed, deleted, or copied. With a traceable database this is not allowed. But it is possible to rename and delete templates and empty folders.

Query Tab

LightCycler® 480 Software includes a query tool you can use to retrieve experiments and other objects stored in the LightCycler® 480 Software database. The query tool is accessible via the Navigator in the form of the *Query* tab.

11. If your LC480 instrument is a new one, this subfolder provides templates for you to edit and kick start your own experiments.
12. If an error message is displayed stating that the query engine needs to be updated, you must update the database. If you have Local Administrator privileges, *see* "Updating the database" in section Administrative Tools for instructions. Otherwise, see your system administrator.

13. The *Sample Selector* and the *Sample Table* are displayed in many windows (*example:* most windows connected to analyses) in the LC480 Software. The *Sample Selector* includes a MWP image with selectable wells and a legend showing selectable sample groups where required. The MWP image can be used to select samples, or as a visual display. When used to select samples, it may appear with or without the legend and may also appear with or without a *Sample Table*. Samples in the MWP image can be enabled or disabled by choosing a subset in the *Subset* combo box. A disabled sample is colored dark gray, exhibits no response when clicked, and shows no information. Samples in the MWP that do not belong to the subset chosen for analysis are disabled by default and cannot be changed. Which sample groups are available in the legends depends on the analysis type. When enabled, a sample may be either selected or deselected. A selected sample is displayed as a pressed button with a white background. A button for a deselected sample is displayed as not pressed with a light blue background. Only selected samples are displayed in the *Results* table and on the corresponding analysis chart.
14. The sample color in the Sample Table always refers to the color in a chart or data display and to the color in the MWP image. Only samples that are enabled and selected in the *Sample Selector* are displayed in the *Sample Table*. Other information (in additional columns) may be added to a *Sample Table* according to the context of the screen (e.g., results such as Cp and concentration for quantification analyses). If there are more samples than can be displayed, a scroll bar is added. The *Sample Selector* and the *Sample Table* are displayed on many windows (example: most windows connected to analyses) in the LightCycler® 480 Software and are used to select the samples to be displayed in the corresponding analysis charts or to include or exclude samples from analysis. For more information on the *Sample Selector*, see the previous section. The Sample Table displays the well coordinates of the samples in the MWP image and the color which represents a sample in the analysis charts (defined by the sample preferences).
15. After you have changed the include status of a sample, you must always recalculate the analysis.
16. Exporting a file does not remove the object from the database, but instead copies the file and stores it at the location you designate. The exported file has an .ixo file extension. You can also export any object in XML format as a Summary XML file. You can export the following objects:

    - User default preferences and user preferences for charts and samples.
    - LightCycler® 480 experiments.
    - Standard melting curve.

- Standard curve object.
- Templates.
- Macros.
- Color Compensation objects.
- LightCycler® 480 Instrument Error and Operation logs.

17. The Experiment Text File format is tab-delimited. It includes two header lines:

    - The first header line contains the experiment name.
    - The second header line contains column headers.

    The data file contains the following information:

    - Sample position.
    - Sample name.
    - Program number.
    - Segment number.
    - Cycle number.
    - Acquisition time.
    - Acquisition temperature.
    - Fluorescence data for each channel.

18. As the experiment progresses, the *Messages* area displays messages indicating information, warnings, and errors encountered during the run. Returned sample data is displayed in the charts on the *Data* tab.

19. Directly after completion of a run, the MWP loader may be hot enough to cause an immediate burn. Wait an appropriate time period to let the loader cool down. Always include a final cooling step in your LightCycler® 480 Instrument run protocol. Be aware that in case of a long standby the MWP itself may be heated to +60 to +80 °C by the heated lid, even if you have cooled down the LightCycler® 480 Instrument to +40 °C after the PCR.

20. During a run, temporary backup data for the current experiment is saved to the user's file system. If the run finishes and has saved the data in the database without an error, these temporary backup data are deleted. If the connection between the application and the instrument is temporarily interrupted, the software will download data automatically from the instrument after the connection is reestablished. The maximum length of a timeout is 7 min. If the timeout is exceeded, the run is considered prematurely terminated, and a warning is generated. If backup or instrument data exist, the data will be automatically recovered upon your next login or start of a new run if a corresponding experiment is found by the software. If no corresponding

experiment is found, the software prompts you to confirm the deletion of the data.

21. You can enter or modify the sample information at any time before, during, or after the experiment is completed. We recommend that you enter the sample information before running the experiment.

22. You cannot make changes to an analysis subset after an analysis is created using the subset.

23. LightCycler® 480 Software includes a batch export tool that lets you export all analysis results from experiments in a group of folders and all its subfolders by using the analysis batch result wizard. Follow the instructions below to export an experiment analysis. A result batch export exports all results of a selected analysis type from all experiments in a group of selected folders or experiments and all of the subfolders at the same time. The results are exported to a single tab-delimited file. Analysis result batch export is only possible from the Navigator. Analysis result batch export is performed using a wizard. You can move from a wizard step to the previous or next step by clicking the corresponding button. Note that the Next button will only be available when you have provided the settings required for the current tab.

24. Unlike some other real-time PCR machines that are able to display the signals from all detection channels simultaneously, the LC480 user cannot view more than one detection channel at a time, example; FAM (533 nm) channel at the same time as the LCR610 (610 nm) channel. For instance, some other machines including the ABI 7500 real-time PCR system [10] allow the user to view the signals from a selection of one, two, or all three channels. Each manufacturer of real-time PCR machines has some advantages over the other. No one machine manufacturer merits all the advantages.

References

1. Vu BN, Jafari AJ, Aardema M, Tran HK, Nguyen DN, Dao TT, Nguyen TV, Tran TK, Nguyen CK, Fox A, Bañuls AL, Thwaites G, Nguyen KV, Wertheim HF (2016) Population structure of colonizing and invasive *Staphylococcus aureus* strains in northern Vietnam. J Med Microbiol 65(pt 4):298–305. doi:10.1099/jmm.0.000220

2. Kolawole DO, Adeyanju Al, Schaumburg F, Akinyoola AL, Lawal OO, Amusa YB, Köck R, Becker K (2013) Characterization of colonizing *Staphylococcus aureus* isolated from surgical wards' patients in a Nigerian university hospital. PLoS One 8(7):e68721. doi:10.1371/journal.pone.0068721

3. Ubukata K, Nakagami S, Nitta A, Yamane A, Kawakami S, Sugiura M, Konno M (1992) Rapid detection of the *mecA* gene in methicillin-resistant staphylococci by enzymatic detection of polymerase chain reaction products. J Clin Microbiol 30:1728–1733

4. Murakami K, Nomura K, Doi M, Yoshida T (1989) Production of low-affinity penicillin-binding protein by low- and high-resistance groups of methicillin-resistant *Staphylococcus aureus*. Antimicrob Agents Chemother 32:1307–1311

5. Foster TJ, Geoghegan JA, Ganesh VK, Hook M (2013) Adhesion, invasion and evasion: the

many functions of the surface proteins of *Staphylococcus aureus*. Nat Rev Microbiol 12(1):49–62

6. Chambers HF (2005) Community-associated MRSA—resistance and virulence converge. N Engl J Med 352:1485–1487
7. David, M.Z., and Daum, R.S (2010) Community-associated methicillin-resistant *Staphylococcus aureus*: epidemiology and clinical consequences of an emerging epidemic. Clin Microbiol Rev 23(3), 616–687.
8. Geha DJ, Uhl JR, Gustaferro CA, Persing DH (1994) Multiplex PCR for identification of methicillin-resistant staphylococci in the clinical laboratory. J Clin Microbiol 32(7): 1768–1772
9. Okolie CE, Wooldridge KG, Turner DP, Cockayne A, James R (2015a) Development of a heptaplex PCR assay for identification of *Staphylococcus aureus* and CoNS with simultaneous detection of virulence and antibiotic resistance genes. BMC Microbiol 15:157
10. RocheAppliedScience.TheLightCycler®480real-time PCR system catalogue. https://lifescience.roche.com/wcsstore/RASCatalogAssetStore/Articles/05237645001_10.12.pdf. Accessed 4 Apr 2016
11. Okolie CE (2009) Development of diagnostic and therapeutic tools for *Staphylococcus aureus* infections. Ph.D. Thesis, University of Nottingham. http://ethos.bl.uk/OrderDetails.do?uin=uk.bl.ethos.517830
12. Okolie CE, Wooldridge KG, Turner DP, Cockayne A, James R (2015b) Development of a new pentaplex real-time PCR assay for the identification of poly-microbial specimens containing *Staphylococcus aureus* and other staphylococci, with simultaneous detection of staphylococcal virulence and methicillin resistance markers. Mol Cell Probes 29(3):144–150
13. Johnsson DP, Stralin MK, Soderquist B (2004) Detection of Panton-valentine leukocidin gene in *Staphylococcus aureus* by LightCycler PCR: clinical and epidemiological aspects. Clin Microbiol Infect 10:884–889
14. Louie L, Goodfellow J, Mathieu P, Glatt A, Louie M, Simor AE (2002) Rapid detection of methicillin-resistant staphylococci from blood culture bottles by using a multiplex PCR assay. J Clin Microbiol 40(8):2786–2790

Chapter 13

Multiplex Peptide Nucleic Acid Fluorescence In Situ Hybridization (PNA-FISH) for Diagnosis of Bacterial Vaginosis

Antonio Machado and Nuno Cerca

Abstract

Fluorescence in situ hybridization (FISH) is a molecular method used to identify and quantify microorganisms in a wide range of samples. This technique combines the simplicity of microscopic observation and the specificity of DNA/rRNA hybridization, allowing detection of selected bacterial species and morphologic visualization. Here, we describe a quantitative molecular diagnosis of bacterial vaginosis, based on the classical Nugent score. Our probes are able to differentiate *Lactobacillus* spp. and *Gardnerella vaginalis* from the other undefined bacterial species considered in the Nugent score.

Key words Fluorescence in situ hybridization (FISH), *Lactobacillus* spp., *Gardnerella vaginalis*, Vaginal samples, Culture cell line, Bacterial vaginosis

1 Introduction

Bacterial vaginosis (BV) is one of the most common perturbations of the normal vaginal condition in women of reproductive age [1] and often exhibits high prevalence, high relapse rates, and associated complications [1, 2], which renders this infection of global importance. BV is associated with increased taxonomic richness and diversity and is normally characterized by a decrease in vaginal lactobacilli and a simultaneous increase in the anaerobes population [2]. Therefore, vaginal bacterial communities differ dramatically between healthy patients and patients with BV, where *Gardnerella vaginalis* is present in over 90% of BV cases [3]. Recent evidence suggests that the presence of a multi-species biofilm, mainly composed of *G. vaginalis* biofilms, is responsible for BV development [3, 4].

The most frequently used method for BV diagnosis is the physician's assessment by Amsel clinical criteria [5]. This method

Kimberly A. Bishop-Lilly (ed.), *Diagnostic Bacteriology: Methods and Protocols*, Methods in Molecular Biology, vol. 1616, DOI 10.1007/978-1-4939-7037-7_13, © Springer Science+Business Media LLC 2017

is fairly subjective and is based on the observation of the following criteria: vaginal fluid pH above 4.5; positive "whiff test" (detection of fishy odor upon 10% potassium hydrogen addition); presence of clue cells (vaginal epithelial cells covered by bacteria) on microscopic examination of vaginal fluid; and homogeneous milky vaginal discharge. At least three from the four clinical signs must be present to establish a positive BV diagnosis [5]. However, the Amsel criteria has been shown to have a low specificity and, therefore, it is not the most appropriate method to diagnose BV [6].

The golden laboratory standard for BV diagnostic is the Nugent score analysis, a microscopic method that quantifies three different bacteria morphotypes presented in the vaginal smears [7]. This analysis uses a Gram stain scoring system based on the evaluation of the following morphotypes: large gram-positive rods (*Lactobacillus* spp. morphotypes); small gram-variable rods (*G. vaginalis* morphotypes); small gram-negative rods (*Bacteroides* spp. morphotypes); and curved gram-variable rods (*Mobiluncus* spp. morphotypes). The vaginal microflora diagnosis is then based on the sum of each morphotype score. Importantly, the evaluation of smears is also subjective and user dependent [8].

Although both methodologies are easy and fast to perform, they do not provide a robust diagnosis of BV. When combined, these standard tests have a sensitivity and specificity of 81 and 70% [5], respectively. To improve BV diagnosis, several new molecular methodologies have been proposed, including fluorescence in situ hybridization (FISH) as an interesting option. This molecular technique combines the simplicity of microscopic observation and the specificity of DNA/rRNA hybridization, allowing the detection of selected bacterial species and morphologic visualization [9]. Originally, FISH was performed using DNA probes. However, new synthetic alternatives have shown to be more advantageous. Peptide Nucleic Acid (PNA) probes improve FISH efficiency because they enable quicker and more specific hybridization [10]. These probes have single bases linked by a neutral peptide backbone, thus avoiding the repulsion between the negatively charged phosphate backbone characteristic of DNA/DNA hybridization [11]. Because PNA is a synthetic molecule, these probes are resistant against cytoplasmic enzymes [12], such as nucleases and proteases, and their hybridization step can be performed efficiently under low salt concentrations [13]. Therefore, these advantages make PNA-FISH a promising tool for the diagnosis of Bacterial vaginosis, providing also a rapid and accurate diagnosis of several other microbial infections [14, 15].

2 Materials

Prepare all solutions using ultrapure water (prepared by purifying deionized water to attain a sensitivity of 18 MΩ cm at 25 °C) or autoclaved deionized water (autoclaved at least 30 min at 120 °C) and analytical grade reagents.

1. Vaginal swabs.
2. Glass slides.
3. Coverslips.
4. Coplin jar.
5. Incubator.
6. Fluorescence microscopy.
7. Non-fluorescent immersion oil.
8. 4% (wt/vol) paraformaldehyde at pH 7.5 (*see* **Note 1**).
9. 50% (vol/vol) ethanol (*see* **Note 1**).
10. Washing solution (5 mM Tris base, 15 mM NaCl, and 0.1% (vol/vol) triton X-100) at pH 10 (*see* **Note 2**).
11. Hybridization solution (10% (wt/vol) dextran sulfate, 10 mM NaCl, 30% (vol/vol) formamide, 0.1% (wt/vol) sodium pyrophosphate, 0.2% (wt/vol) polyvinylpyrrolidone, 0.2% (wt/vol) ficoll, 5 mM disodium EDTA, 0.1% (vol/vol) triton X-100, and 50 mM Tris–HCl) at pH 7.5 (*see* **Note 3**).
12. Lac663 probe (Alexa Fluor 488-ACATGGAGTTCCACT; HPLC purified >90%; *see* **Notes 4** and **5**).
13. Gard162 probe (Alexa Fluor 594-CAGCATTACCACCCG; HPLC purified >90%; *see* **Notes 4** and **5**).
14. 4′,6-diamidino-2-phenylindole (DAPI) stain (*see* **Note 5**).
15. Saline solution (0.9% NaCl prepared in distilled water).
16. Phosphate buffer saline (PBS) solution.

3 Methods

Carry out all procedures at room temperature unless otherwise specified.

3.1 Preparation of FISH Samples

All vaginal samples must be prepared for further FISH analysis, more precisely, these samples must be purified and diluted for an optimal hybridization procedure and microscopic visualization.

3.1.1 Preparation of Vaginal Samples

1. Collect vaginal swabs after being brushed against the lateral vagina wall (*see* **Note 6**) and place each vaginal swab in a culture swab transport media (for example, the culture swab transport system VWR CE0344).

2. Transport the swabs to the laboratory facility as soon as possible (*see* **Note 7**).
3. Next, immerse each swab into a microtube with 1 mL of sterile PBS and mix by vortexing.
4. Centrifuge each microtube at 17,000 × *g* during 5 min at room temperature.
5. Afterward, disperse the obtained supernatant and resuspend the pellet in 1 mL of saline solution (0.9% NaCl prepared in distilled water).
6. Dilute 100 μL of each resuspended pellet into 900 μL of saline solution or PBS (dilution of 1:10) to eliminate possible contaminants for FISH procedures.
7. For each diluted vaginal sample, spread 20 μL of the final suspension on a glass slide and dry the glass slide at room temperature or in an incubator at 37 °C during 10 min.

3.2 FISH Hybridization Procedure

For each air-dried glass slide, spread 50–100 μL of the solution of paraformaldehyde 4% (w/v) on the glass slide and dry them at room temperature during 10 min (*see* **Note 8**).

1. Remove the remaining paraformaldehyde 4% (w/v) and spread again 50–100 μL of the solution of ethanol 50% (v/v) on the glass slide (*see* **Note 8**). Then, dry the slide at room temperature during 10 min.
2. After the fixation step (*see* **Note 9**), cover the samples with 20 μL of hybridization solution containing 200 nM of each PNA probe (Lac663 and Gard162 PNA probe aliquots).
3. Subsequently, cover the samples on the glass slides with coverslips and incubate them in humidified chambers at the hybridization temperature of 60 °C for 90 min.
4. Next, remove the coverslips and immerse the slides into a coplin jar with a pre-warmed washing solution at 60 °C for 30 min (*see* **Note 10**).
5. Remove the glass slides from the coplin jar and allow them to dry at room temperature.
6. Then, perform an additional 4′,6-diamidino-2-phenylindole (DAPI) staining step by covering each surface of the glass slides with 10 μL of DAPI for 5 min at room temperature in the dark (*see* **Note 11**).
7. Wash the DAPI stained glass slides with 10 μL of sterile PBS and repeat twice this washing step.
8. Immediately, observe the stained glass slides in the fluorescence microscope (*see* Subheading 3.3).

3.3 Microscopic Visualization

Prior to microscopic visualization, turn on the fluorescence light on the microscopy at least for 15 min (*see* **Note 5**).

1. Add one drop of non-fluorescent immersion oil to each glass slide and cover them with coverslips.
2. Focus the epifluorescence microscope first with an empty glass slide with the objective of 4×, 40×, and finally 100×.
3. Connect the microscope's camera and evaluate each filter's working ability for each PNA probe or fluorochrome used in the previous FISH procedure (*see* **Note 12**).
4. From each glass slide and each fluorescence filter (*see* **Note 12**), take 20 images from random regions using a total magnification of ×1000.

Two samples of FISH illustrating results obtained with vaginal samples are shown in Fig. 1.

3.4 Bacteria Count and BV Diagnosis

The bacterial load of the vaginal samples must be evaluated based on the Nugent criteria score (*see* Table 1).

1. For each sample, collect at least 20 images taken by each filter (blue, green, and red) and count the total number of bacteria in each image.

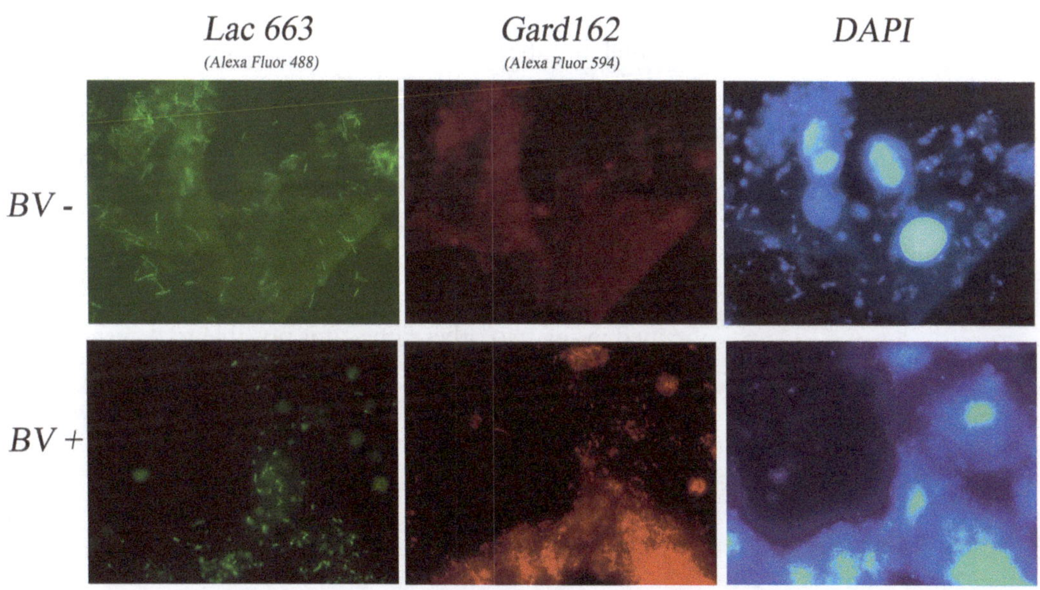

Fig. 1 Fluorescence microscopy pictures of *Lactobacillus* spp., *Gardnerella vaginalis*, and other bacterial species from healthy and BV vaginal clinical samples by specific PNA probes (Lac663 and Gard162) associated with Alexa Fluor 488 and 594 fluorochromes and DAPI staining, respectively. Both vaginal swab samples exhibited a different vaginal microflora profile. As shown in the *green filter*, healthy and BV samples showed the presence of *Lactobacillus* spp. species but only BV sample demonstrated an elevated *G. vaginalis* concentration in the vaginal swabs (*red filter*), which proved to establish clue cells on the vicinity of b vaginal epithelial cells (*blue filter*)

Table 1
Scheme for grading vaginal contents based on Nugent score system, adapted from Nugent et al. [7]

Nugent'sscoring system			
Score	*Lactobacillus* spp. morphotype	*Gardnerella* spp. morphotype	*Mobiluncus* spp. morphotype
0	4+	0	0
1	3+	1+	1+ or 2+
2	2+	2+	3+ or 4+
3	1+	3+	–
4	0	4+	–
Vaginal microflora diagnosis by Nugent score system			
Total score		**Interpretation**	
0–3		Normal vaginal microflora	
4–6		Intermediate vaginal microflora	
7–10		Bacterial vaginosis in vaginal microflora	

Morphotypes are scored as the average number per oil immersion field. Quantification of each individual score: 0 for no morphotype present; 1+ for 1 morphotype present; 2+, 1–4 morphotypes present; 3+, 5–30 morphotypes present; 4+, 30 or more morphotypes present. Total score is the sum of the average classification of *Lactobacillus*, *Gardnerella*, and finally *Mobiluncus* spp.

2. The total number of bacteria detected specifically with the green filter represents the *Lactobacillus* species (Lac663 PNA probe).
3. The total number of bacteria detected specifically with the red filter represents *Gardnerella vaginalis* species (Gard162 PNA probe).
4. The total number of bacteria from the blue filter (DAPI) represents the total bacterial load. By subtracting the values obtained in the two other filters, it is possible to determine the total number of other bacteria, such as *Mobiluncus* spp.
5. Next, score each sample's morphotypes (*Lactobacillus* spp., *G. vaginalis* and other bacteria) as the average number, see per filter field (*see* **Note 13**).
6. Then, calculate the total score, for each sample, by the sum of the average classification of each morphotype (*see* Table 1).

4 Notes

1. The optimal fixation solutions must be evaluated of each specific PNA probe and depend on the sample complexity. The optimal fixation solutions for BV diagnosis with the actual PNA probes are paraformaldehyde 4% (w/v) and ethanol 50% [16].

These fixation solutions can be preserved up to 2 years in storage: the paraformaldehyde 4% must be stored at −20 °C while the ethanol 50% can be stored at 4 °C.

Solution of paraformaldehyde 4%

- For a final volume of 100 mL, add 4.0 g of paraformaldehyde to 95 mL of sterile water, allowing the magnetic stir bar to spin slowly.
- Adjust the pH to 7.5 with HCl 1 M and add sterile water to a final volume of 100 mL.
- Filter the final solution through a nitrocellulose membrane of 0.45 μm and store at −20 °C until 30 min before starting the hybridization protocol.
- If using a glass beaker, paraformaldehyde can be dissolved faster by increasing the water temperature to 60 °C and allowing the magnetic stir bar to spin faster.

Solution of ethanol 50%

- For a final volume of 100 mL, add 50 mL of absolute ethanol to 50 mL of sterile water, allowing the magnetic stir bar to spin slowly for 2 min.
- Store at 4 °C until 30 min before starting the hybridization protocol.

2. The washing solution must be prepared less than 48 h before the hybridization procedure and must be stored at 4 °C until the experiment. This solution consisted of 5 mM Tris base, 15 mM NaCl, and 0.1% (vol/vol) triton X-100 (at pH 10).
   - For a final volume of 500 mL, add 0.303 g of Tris Base and 0.438 g of NaCl to 250 mL of sterile water. Next, mix and homogenize the previous amounts in the initial volume of 250 mL.
   - Next, add 500 μL of Triton X-100 and homogenize again this solution.
   - Add sterile water to a volume of 495 mL and adjust the pH with NaOH 1 M (*see* **Note 2**).
   - Make up to 500 mL with sterile water and autoclaved the solution at 120 °C for 20 min.
   - Store at 4 °C until 30 min before starting the hybridization protocol.
   - The washing solution must have a pH of 10; usually, it is only required to add about 200 μL of NaOH 1 M solution in the washing solution to achieve the desired pH value.
3. The hybridization solution must have a pH of 7.5 and must be stored at 4 °C until the experiment. The hybridization solution can be preserved until 1 year in storage. The hybridization

solution contains 10% (wt/vol) dextran sulfate, 10 mM NaCl, 30% (vol/vol) formamide, 0.1% (wt/vol) sodium pyrophosphate, 0.2% (wt/vol) polyvinylpyrrolidone, 0.2% (wt/vol) FICOLL, 5 mM disodium EDTA, 0.1% (vol/vol) triton X-100 and 50 mM Tris–HCl.

- For a final volume of 10 mL, add 1.0 g of dextran sulfate, 0.0058 g of NaCl, 0.01 g of sodium pyrophosphate, 0.02 g polyvinylpirroline, 0.02 g FICOLL, 0.02 g disodium EDTA, and 0.079 g of Tris–HCl to 5 mL of sterile water. Next, mix and homogenize the previous amounts in the initial volume of 5 mL.
- Next, add 3.0 mL of formamide and 10 μL of Triton X-100 and mix again this solution.
- Add sterile water to a volume of 9 mL and adjust the pH with HCl 1 M. The hybridization solution must have a pH of 7.5, which helps to dissolve the dextran sulfate.
- Make up to 10 mL with sterile water and filter through a nitrocellulose membrane of 0.45 μm.
- Store at 4 °C until 30 min before starting the hybridization protocol.

4. The PNA probe aliquots are prepared by a dilution of the original and commercial PNA probe with the previous hybridization solution and must be stored in the dark at 4 °C until the experiment. The PNA probe aliquots can be preserved until 1 year in storage.

   For Bacterial vaginosis diagnosis by PNA-FISH, the selected PNA probes are Lac663 and Gard162 probes. As previously described by Machado et al. [16], the selected sequences were synthesized by Panagene (Daejeon, South Korea) and the oligonucleotides N terminus was attached to an Alexa Fluor molecule via a double 8-amino-3,6-dioxaoctanoic acid (AEEA) linker. The Gard162 probe hybridizes between positions 162 and 176 of the *G. vaginalis* strain 409–05 16S rRNA sequence (RDPII ID: S001872672) and was selected for probe design (PNA Probe: Gard162, Alexa Fluor 594-OO-CAGCATTACCACCCG; HPLC purified >90%). For the detection of *Lactobacillus* spp., Lac663 probe was chosen [16]. This probe was attached to an Alexa Fluor 488 molecule, also via an AEEA linker (PNA Probe: Lac663, Alexa Fluor 488-OO-ACATGGAGTTCCACT; HPLC purified >90%).

   - Add 10 μL of the original PNA probe to 990 μL of the previous hybridization solution, obtaining the desired 200 nM of PNA probe aliquots.
   - Next, cover the microtube of the PNA probe aliquot with aluminum foil and store at 4 °C until the experiment.

5. For microscopic visualization, any standard epifluorescence microscope equipped with a photographic camera and filters capable of detecting the two PNA probes and the 4′,6-diamidino-2-phenylindole (DAPI) stain can be used. The Lac663 probe is attached to the Alexa Fluor 488 molecule, and therefore it is necessary to possess an excitation filter of 470–490 nm with a barrier filter of 516 nm in the microscope. In addition, the Gard162 probe is attached to the Alexa Fluor 594 molecule and it is also necessary that the microscope have an excitation filter of 530–550 nm with a barrier filter of 591 nm. Finally, a narrow excitation band and longpass emission filter could be used as filter for the unspecific fluorochrome DAPI, such as an excitation filter of 365–370 nm with a barrier filter of 421 nm in the microscope.
6. The vaginal swabs must be brushed against the lateral vaginal wall to collect the vaginal fluid sample.
7. The referred vaginal swabs instructions recommend the transportation of the vaginal fluid sample in the first 48 h to guarantee the bacteria isolation and growth in media. However, it is still possible to perform extraction and hybridization of the recollected bacteria up to 3 weeks. This timeline strictly depends on the selected vaginal swabs kit used in our work.
8. The surface containing the vaginal bacteria on the glass slide must be covered by the fixation solutions; therefore, the applied fixation volume can vary between 50 and 100 μL.
9. After the fixation step, the researchers or technicians must disconnect the lights of the laboratory and carry on the hybridization protocol in the dark or with few illumination, avoiding burnout of the fluorescence from the PNA probe aliquots. Also, the fixated samples can be stored in the dark for 12–18 h at 4 °C, before continuing the hybridization step and microscopic visualization.
10. A coplin jar filled with washing solution must be pre-warmed at 60 °C for 30–60 min before starting the washing step on the hybridized samples.
11. The concentration of DAPI staining solution depends on the commercial product acquired; therefore, it is advised to follow the manufacturer's instructions. An optimization step might be needed, in order to balance the fluorophore intensities of the three dyes.
12. In each experimental assay and for each sample, 20 images of each filter (blue, green, and red) must be taken from the same random regions. It is advised to start recollecting images from lesser fluorescent dye to the most fluorescent dye, avoiding consumption of the other fluorescence probes by fluorescence light of the microscope.

13. The Nugent score system is based on the total score obtained by the sum of the average classification of the following morphotypes: *Lactobacillus*, *Gardnerella* and *Bacteroides*, and finally *Mobiluncus* spp. [7]. However, the actual PNA-FISH methodology only classified the *Lactobacillus* spp. morphotype through Lac663 probe, *Gardnerella* spp. through Gard162 probe, and Mobiluncus spp. morphotype through the remaining bacteria stained with DAPI [17].

Acknowledgment

Research on BV biofilms in NC laboratory is supported by funding from the Fundação para a Ciência e a Tecnologia (FCT) strategic project of unit UID/BIO/04469/2013.

References

1. Bretelle F, Rozenberg P, Pascal A, Favre R, Bohec C, Loundou A, Senat MV, Aissi G, Lesavre N, Brunet J, Heckenroth H, Luton D, Raoult D, Fenollar F, Groupe de Recherche en Obstetrique Gynecologie (2015) High *Atopobium vaginae* and *Gardnerella vaginalis* vaginal loads are associated with preterm birth. Clin Infect Dis 60:860–867
2. Tibaldi C, Cappello N, Latino MA, Masuelli G, Marini S, Benedetto C (2009) Vaginal and endocervical microorganisms in symptomatic and asymptomatic non-pregnant females: risk factors and rates of occurrence. Clin Microbiol Infect 15:670–679
3. Verstraelen H, Swidsinski A (2013) The biofilm in bacterial vaginosis: implications for epidemiology, diagnosis and treatment. Curr Opin Infect Dis 26:86–89
4. Machado A, Cerca N (2015) Influence of biofilm formation by *Gardnerella vaginalis* and other anaerobes on bacterial vaginosis. J Infect Dis 212:1856–1861
5. Forsum U, Hallén A, Larsson P (2005) Bacterial vaginosis-a laboratory and clinical diagnostics enigma. Acta Pathol Microbiol Immunol Scand 113:153–161
6. Money D (2005) The laboratory diagnosis of bacterial vaginosis. Can J Infect Dis Med Microbiol 16:77–79
7. Nugent R, Krohn M, Hillier S (1991) Reliability of diagnosing bacterial vaginosis is improved by a standardized method of gram stain interpretation. J Clin Microbiol 29:297–301
8. Sha BE, Chen HY, Wang QJ, Zariffard MR, Cohen MH, Spear GT (2005) Utility of Amsel criteria, Nugent score, and quantitative PCR for Gardnerella vaginalis, Mycoplasma hominis, and Lactobacillus spp. for diagnosis of bacterial vaginosis in human immunodeficiency virus-infected women. J Clin Microbiol 43:4607–4612
9. Justé A, Thomma BP, Lievens B (2008) Recent advances in molecular techniques to study microbial communities in food-associated matrices and processes. Food Microbiol 25:745–761
10. Peleg AY, Tilahun Y, Fiandaca MJ, D'Agata EMC, Venkataraman L, Moellering RC, Eliopoulos GM (2009) Utility of peptide nucleic acid fluorescence *in situ* hybridization for rapid detection of *Acinetobacter* spp. and *Pseudomonas aeruginosa*. J Clin Microbiol 47:830–832
11. Stender H, Fiandaca M, Hyldig-Nielsen JJ, Coull J (2002) PNA for rapid microbiology. J Microbiol Methods 48:1–17
12. Amann R, Fuchs BM (2008) Single-cell identification in microbial communities by improved fluorescence in situ hybridization techniques. Nat Rev Microbiol 6:339–348
13. Almeida C, Azevedo NF, Iversen C, Fanning S, Keevil CW, Vieira MJ (2009) Development and application of a novel peptide nucleic acid probe for the specific detection of *Cronobacter* genomospecies (*Enterobacter sakazakii*) in powdered infant formula. Appl Environ Microbiol 75:2925–2930

14. Shepard JR, Addison RM, Alexander BD, Della-Latta P, Gherna M, Haase G, Hall G, Johnson JK, Merz WG, Peltroche-Llacsahuanga H, Stender H, Venezia RA, Wilson D, Procop GW, Wu F, Fiandaca MJ (2008) Multicenter evaluation of the *Candida albicans/Candida glabrata* peptide nucleic acid fluorescent *in situ* hybridization method for simultaneous dual-color identification of *C. albicans* and *C. glabrata* directly from blood culture bottles. J Clin Microbiol 46:50–55
15. Trnovsky J, Merz W, Della-Latta P, Wu F, Arendrup MC, Stender H (2008) Rapid and accurate identification of *Candida albicans* isolates by use of PNA FISH flow. J Clin Microbiol 46:1537–1540
16. Machado A, Almeida C, Salgueiro D, Henriques A, Vaneechoutte M, Haesebrouck F, Vieira MJ, Rodrigues L, Azevedo NF, Cerca N (2013) Fluorescence *in situ* hybridization method using peptide nucleic acid probes for rapid detection of *Lactobacillus* and *Gardnerella* spp. BMC Microbiol 13:82
17. Machado A, Castro J, Cereija T, Almeida C, Cerca N (2015) Diagnosis of bacterial vaginosis by a new multiplex peptide nucleic acid fluorescence *in situ* hybridization method. Peer J 3:e780

Chapter 14

A Closed-tube Loop-Mediated Isothermal Amplification Assay for the Visual Endpoint Detection of *Brucella* spp. and *Mycobacterium avium* subsp. *paratuberculosis*

Marcos D. Trangoni, Andrea K. Gioffré, and Silvio L. Cravero

Abstract

LAMP (loop-mediated isothermal amplification) is an isothermal nucleic acid amplification technique that is characterized by its efficiency, rapidity, high yield of final product, robustness, sensitivity, and specificity, with the blueprint that it can be implemented in laboratories of low technological complexity. Despite the conceptual complexity underlying the mechanistic basis for the nucleic acid amplification, the technique is simple to use and the amplification and detection can be carried out in just one step. In this chapter, we present a protocol based on LAMP for the rapid identification of isolates of *Brucella* spp. and *Mycobacterium avium* subsp. *paratuberculosis*, two major bacterial pathogens in veterinary medicine.

Key words Lamp, Isothermal amplification, Molecular detection, *Brucella*, *Mycobacterium avium* subsp. *paratuberculosis*, Brucellosis, Paratuberculosis

1 Introduction

Brucella spp. and *Mycobacterium avium* subsp. *paratuberculosis* (MAP) are pathogenic microorganisms of veterinary concern. *Brucella* spp. are the etiological agents of brucellosis, leading to abortion in cattle, sheep, pigs, and goats, while MAP is the causal agent of Johne's disease in cattle, a chronic diarrhea and wasting disease [1, 2]. The traditional methods for detecting these pathogens are largely based on phenotypic traits, and the diagnosis typically involves bacteriological culture, histopathology, and serological tests [3, 4]; however, the isolation by culture is considered the gold standard. This process is time consuming for MAP, which requires up to 2 months' growth in culture media. Nucleic acid amplification has allowed the sensitive diagnosis of different bacteria, minimizing the requirement of biosafety conditions and often replacing time-consuming techniques. In addition to contributing to the diagnosis, nucleic acid amplification provides an accurate

Kimberly A. Bishop-Lilly (ed.), *Diagnostic Bacteriology: Methods and Protocols*, Methods in Molecular Biology, vol. 1616, DOI 10.1007/978-1-4939-7037-7_14,

molecular tool for identification at the species or subspecies level. The loop-mediated isothermal amplification (LAMP) technique is characterized by its simplicity because the entire process of amplification and detection can be performed in a single step [5, 6]. The *Bst* polymerase plays a key role in the LAMP reaction process; in addition to its polymerization activity, its displacement activity can separate the non-template strand from the template DNA under isothermal condition, and so this technology requires less specialized equipment than conventional PCR. Compared to PCR, the LAMP assay displays an equivalent sensitivity and specificity, but the reaction time shorter. Thus, LAMP is an interesting and promising option for rapid pathogen identification and diagnosis.

In previously published work, we developed two protocols based on conventional targets *bscp*31 and IS*900* to identify *Brucella* spp. and MAP, respectively. Guidelines are presented herein to help ensure the simple, rapid, and specific detection of *Brucella* spp. and MAP with the same sensitivity as conventional PCR.

2 Materials

2.1 General Considerations

All buffers and solutions should be prepared with ultrapure, molecular biology grade water. Plasticware used must be RNase/DNase-free certified. Disposable filter tips must be used for all pipetting. In order to prevent contamination with target DNA, the preparation of the reagents and the steps involving DNA manipulation must be performed in separate rooms, and at the end of the amplification reaction it is mandatory to avoid opening tubes in the area where the reaction is set up. Separate sets of pipettors and tips should be used specifically for manipulations involving DNA or DNA-free manipulations in order to preclude contaminating reagents. LAMP reaction mixtures should be prepared on ice.

2.2 Equipment

1. Biological safety cabinet.
2. Dry block heater or water bath.
3. Benchtop microcentrifuge.
4. Disposable plasticware: sterile 0.2-, 0.5-, and 1.5-mL microtubes and racks; RNase/DNase-free sterile filter tips (P10, P20, P200, P1000); sterilized inoculating loops (for 10 μL inoculation).
5. Two sets of pipettes covering the volumes 200–1000; 20–200; 2–20; 1–10 μL.

2.3 Reagents and Solutions

1. Ultrapure, molecular biology grade water.
2. Oligos Buffer: 10 mM Tris–HCl pH 8, 1 mM EDTA.
3. LAMP primers according the pathogen to test (*see* Table 1 and **Note 1**).

Table 1
List of LAMP primers

Organism (target)	Temperature	Primer name	Sequence (5′–3′)	Reference
Brucella spp. (*bcsp*31)	60 °C	F3-Bru	CAGACGTTGCCTATTGGGC	[7]
		B3-Bru	GGCTCATCCAGCGAAACG	
		FIP-Bru	CGGGTAAAGCGTCGCCAGAAGTTTT-GCACCGGCCTTTATGATGG	
		BIP-Bru	ACGATCCATATCGTTGCGCGTTTTT-GCTTGCCTTTCAGGTCTGC	
		LF-Bru	CGCAAATCTTCCACCTTGCC	
		LR-Bru	GGATGCAAACATCAAATCGGTC	
M. avium subsp. *paratuberculosis* (IS*900*)	65 °C	F3-MAP	CGCAACGCCGATACCGT	[7]
		B3-MAP	CCCAGGATGACGCCGAA	
		FIP-MAP	CATCACCTCCTTGGCCAGGC-CCGCTAACGCCCAACAC	
		BIP-MAP	GCGACACCGACGCGATGAT-TCCGGGCATGCTCAGGA	
		LF-MAP	AGTGGCCGCCAGTTGTTG	
		LR-MAP	ACCGCCACGCCGAAATC	

4. 1 M and 10 mM Tris–HCl, pH 8.
5. 25 mM NaOH.
6. SYBR Green I nucleic acid stain, 10,000× concentrate in DMSO. To prepare the working solution (1:10 in water), mix 50 μL of commercially available 10,000× SYBR Green I nucleic acid stain with 450 μL of water. Make aliquots of 50 μL and keep it at −20 °C protected from light until use (*see* **Note 2**).
7. 100 mM $MgSO_4$. Store at −20 °C.
8. 100 mM dNTPs set. Store at −20 °C.
9. 5 M betaine solution (Sigma-Aldrich) (*see* **Note 3**). Store at 4 °C.
10. *Bst* DNA polymerase, large fragment (8 U/μL) (New England Biolabs). Store at −20 °C.
11. 10× ThermoPol buffer: 100 mM $(NH4)_2SO_4$, 100 mM KCl, 200 mM Tris–HCl, 1% Triton X-100, 20 mM $MgSO_4$, pH 8.8 (New England Biolabs). Store at −20 °C.
12. Positive control DNA sample (*see* **Note 4**). Store at −20 °C.

3 Methods

The main steps to perform LAMP for the detection of MAP or *Brucella* spp. are summarized sequentially in Fig. 1.

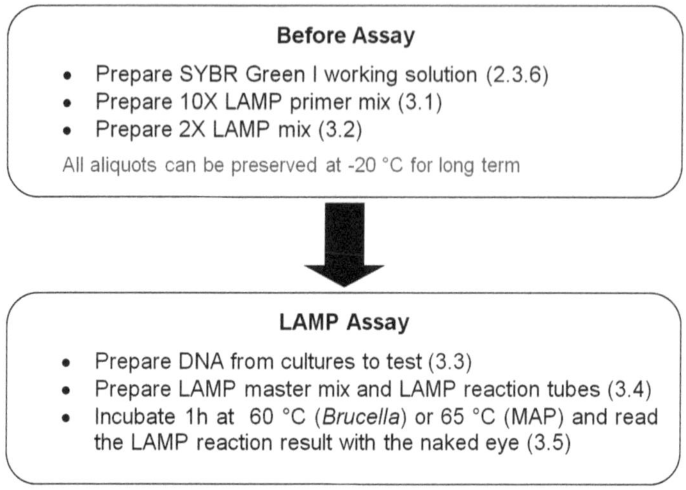

Fig. 1 Sequential steps to perform LAMP to test *Brucella* spp. or MAP from cultures

Table 2
Preparation of 10× LAMP primers mix

Reagent	Stock solution (μM)	[a]Volume (μL)	10× LAMP primers mix (μM)
Oligos buffer	–	48.80	–
Primer F3	100	1.60	1.60
Primer B3	100	1.60	1.60
Primer FIP	100	16.00	16.00
Primer BIP	100	16.00	16.00
Primer LF	100	8.00	8.00
Primer LR	100	8.00	8.00

[a]Volumes listed are for 100 μL of 10× LAMP primers mix

3.1 Preparation of 10× LAMP Primers Mix

1. Reconstitute each primer in Oligos Buffer to get 100 μM stock solutions (*see* **Note 5**).
2. Mix the tubes by inversion and rotation ten times.
3. Spin the tubes for 3 s in a benchtop microcentrifuge and put the tubes on ice.
4. Use the 100 μM primers stock and Oligos Buffer volume indicated in Table 2 to prepare 100 μL of 10× LAMP primers mix (*see* **Note 6**).
5. Mix by pipetting or tapping ten times.
6. Store the 10× LAMP primers mix tubes at −20 °C.

Table 3
Preparation of 2× LAMP mix. List of reagents, initial and final concentrations

Reagent	Stock solution	[a]Volume (μL)	2× LAMP mix
Water		248	
Betaine	5 M	320	1.6 M
dATP	100 mM	28	2.8 mM
dTTP	100 mM	28	2.8 mM
dCTP	100 mM	28	2.8 mM
dGTP	100 mM	28	2.8 mM
$MgSO_4$	100 mM	120	12 mM
ThermoPol buffer	10×	200	2×

[a]Volumes indicated are for 1 mL of 2× LAMP mix

3.2 Preparation of 2× LAMP Mix

1. Thaw all LAMP reagents (except *Bst* DNA polymerase and primers, *see* Table 3) at room temperature and keep on ice.
2. Prepare 2× LAMP mix using the reagents volume indicated in Table 3 (*see* **Note 7**).
3. Make 100 μL aliquots to minimize the number of freeze/thaw cycles.
4. Store the aliquots at −20 °C until use (*see* **Note 8**).

3.3 DNA Extraction from Culture

1. Pick one single colony to test from the corresponding selective media and resuspend in a 1.5 mL microtube containing 50 μL of 25 mM NaOH (*see* **Note 9**).
2. Inactivate the bacterial suspension in a water bath for 5 min at 98–100 °C for *Brucella* spp. or 10 min at 100 °C for MAP (*see* **Note 10**).
3. Add 4 μL of Tris–HCl buffer (1 M, pH 8.0) to neutralize the bacterial suspension.
4. Centrifuge to pellet the unbroken cells at 10,000 × *g* for 5 min at room temperature.
5. Transfer supernatant into a new 1.5 mL microtube for use as DNA template. Add 2 μL to the reaction tube for the LAMP assay (*see* **Note 11**).

3.4 LAMP Operating Procedure

Each isothermal amplification reaction is prepared to a final volume of 25 μL, including the addition of 2 μL of template (*see* **Note 12**).

1. Prepare a master reaction mixture for all the DNA samples to test, including the positive and negative (molecular biology

Table 4
Master Mix reagents volumes for a single LAMP reaction

Reagent	[a]Volume (μL)
Water	7
2× LAMP mix	12.5
10× LAMP primers mix	2.5
Bst polymerase	1
DNA (*see* **Note 15**)	2

[a]Volumes required for 25 μL LAMP reactions

grade water) controls. Set up LAMP master reaction mixture according to Table 4.

2. Add all the reagents except DNA (*see* **Note 13**).
3. Dispense 23 μL of the master mix into each reaction tube (0.2 mL microtubes).
4. Carefully add 1 μL of SYBR Green I working solution inside the cap of each tube (*see* **Note 14** and Fig. 2a).
6. Add 2 μL of each template DNA to be tested, and the same volume of a related template DNA (positive control) or molecular biology grade water (negative control) and gently close the tubes (*see* **Note 15**).

3.5 LAMP Reaction and End Point Detection

1. Incubate in a water bath or a block heater for 60 min at the temperature corresponding to the microorganism to be detected (*see* Table 1).
2. Inactivate the polymerase to stop the reaction at 80 °C for 5 min (*see* **Note 16**).
3. Rehydrate the SYBR Green I which remained in the cap of each reaction tube mixing thoroughly by inversion (*see* Fig. 2b and **Note 17**).
4. Visualize (with the naked eye) the end point of the reaction. The reaction mixture will turn green in the presence of LAMP products, while it will remain orange in their absence (*see* Fig. 2c and **Note 18**).

4 Notes

1. While PCR requires only a pair of primers to amplify target gene sequences, LAMP is quite complex since four to six primers are needed. F3 and B3 are the external primers necessary for initial steps; FIP and BIP are hybrid primers and essential

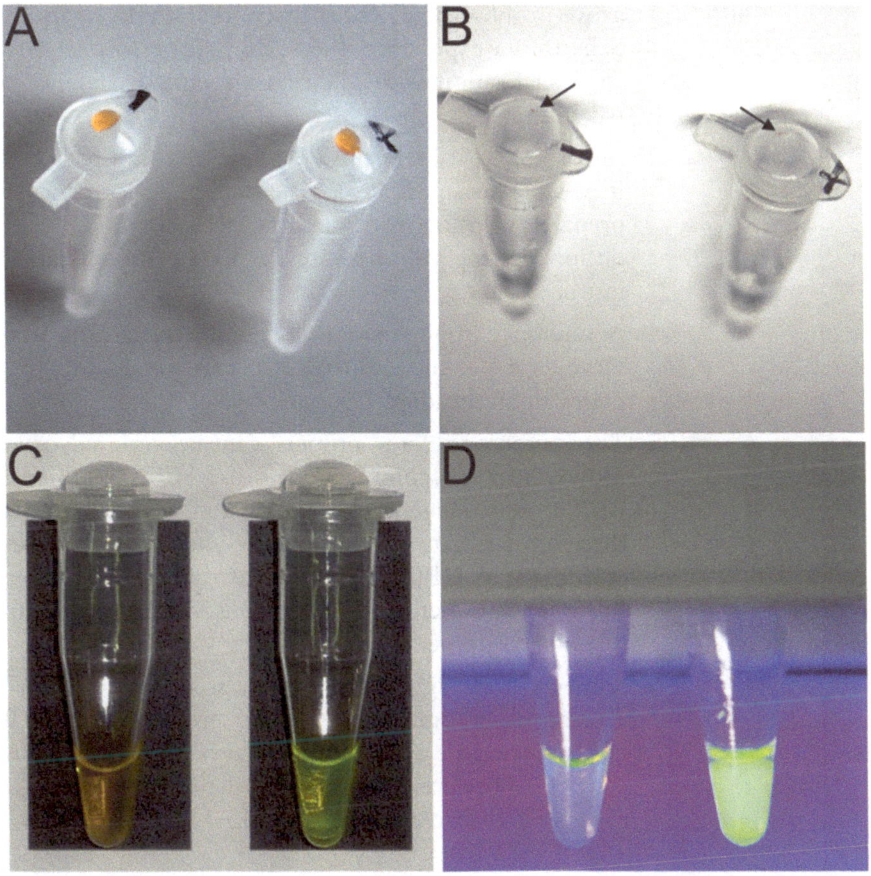

Fig. 2 SYBR Green I for visualization of LAMP reaction results. (**a**) SYBR Green I is added into the cap during the preparation of the LAMP reaction tubes. (**b**) The same tube as (**a**), post-LAMP reaction. The drop has dehydrated and is barely seen (indicated by *arrowheads*). (**c**) Visualization of tubes with the naked eye after rehydration of SYBR Green I. A positive test reaction (*green*) is clearly differentiated from a negative test (*orange*). (**d**) Fluorescence of a positive sample when a UV-transilluminator is used

in all reactions; LF and LR are optional and help in the reaction. The primer design is the bottleneck of the assay and can be carried out with bioinformatics support. In the present chapter, the Bru-LAMP and MAP-LAMP primers were designed with the publicly available software https://primerexplorer.jp/e/v4_manual/index.html.

2. SYBR Green I is commercially available as a concentrated solution in DMSO at 10,000×, used usually in real-time amplifications. In LAMP protocols, a concentration higher than that described for real-time PCR is used. However, the enzyme used for LAMP is also inhibited at high concentrations of SYBR Green I, and therefore the addition of the dye to the reaction mixture is usually performed post-amplification or must be physically separately until amplification reaction is finished,

as we propose here. Another dye that could be used is SYBR Safe. In our experience, the SYBR Safe performance is suitable for visualization with UV-transilluminator but not with the naked eye.

3. According to our results, the use of commercially available 5 M betaine solution leads to better performance than the prepared solution from the solid drug. If you choose this last option betaine chlorhydrate cannot be used as a substitute, just use betaine.
4. A reference strain or a well-characterized field strain must serve as a positive control. Process this strain as explained below in Subheading 3.3.
5. The outer and loop primers can be ordered as desalted. However, it is recommended to use HPLC-purified FIP and BIP primers. Depending on the inner primers and the target, these primers may work as desalted; however, they may not be as efficient as HPLC purified. It is strongly recommended to spin the tubes before primer reconstitution to prevent loss of lyophilized primer mass.
6. Some authors suggest keeping the stock primers at −70 °C to avoid primer degradation. In this protocol, Oligos Buffer is used rather than water; hence, you could store the primers at −20 °C. However, for long-term preservation it is recommended to store each 100 μM LAMP primer stock tube at −70 °C.
7. 2× LAMP mix is a concentrated solution containing all the components required for LAMP reaction, except primers, *Bst* polymerase, and DNA. This mix strongly reduces pipetting steps, increasing throughput and reproducibility, while reducing the risk of contamination, mainly for routine testing.
8. Aliquots of 2× LAMP mix retain functional properties over 2 years preserved at −20 °C.
9. For confluent growth take a loopful. Broth cultures can also be used as a starting point for DNA extraction. For *Brucella* use tryptose agar plates or tryptic soy broth (Difco, BD, USA) and for MAP use conventional fecal culture media (Herrold's egg yolk agar containing mycobactin) or 7H9 liquid medium (Difco, BD, USA) supplemented with 0.2% mycobactin J (Allied Monitor, Fayette, MO, USA).
10. Up to this step you need a biosafety cabinet. Once the sample is inactivated, you could work on a conventional laboratory bench top. If liquid culture medium is used, wash the cells with water as follows. Centrifuge a volume of culture (100–1000 μL depending on growth phase) at 10,000 × *g* for 2 min and remove the supernatant. Then, add 100–200 μL of water

and vortex briefly for washing. Repeat the step twice and resuspend the pellet in 100 μL of ultrapure water or 10 mM Tris–HCl pH 8.

11. The cellular lysate can be repeatedly freeze/thawed.
12. The final reaction volume can be reduced, thus reducing the cost of the reaction. According to our experience, satisfactory results were obtained by using a final volume of 10 μL (with up to 3 μL of template DNA).
13. Similar to PCR, this step, and the following steps, must be performed on ice to avoid nonspecific amplifications.
14. This step must be performed carefully to avoid contact with the master mix. As mentioned above, high concentration of SYBR Green I has an inhibitory effect on *Bst* polymerase.
15. Consider the total number of samples to test, including the positive and negative controls, to prepare the volume of master mix needed. Dispense the volume of master mix, without DNA, in each reaction tube to further add the template DNA. Perform this last step in a separated room with a different pipette to avoid contamination of reagents with DNA.
16. The inactivation step is important to avoid false-positive results. However, this step can be omitted if the visual assessment (Subheading 3.5, **step 3**) is performed quickly after the amplification due to the SYBR Green I inhibitory effect.
17. This step does not require opening the tube, and strongly reduces cross-contamination.
18. The positive and negative controls must be visualized green and orange respectively. A UV-transilluminator or even a domestic use currency reader may be used to increase the fluorescence intensity (*see* Fig. 2d).

References

1. Corbel M (1989) Brucellosis: epidemiology and prevalence worldwide. Brucellosis: clinical and laboratory aspects. CRC Press, Boca Raton, FL, pp 26–37
2. Collins MT (2003) Paratuberculosis: review of present knowledge. Acta Vet Scand 44(3–4): 217–221
3. Gall D, Nielsen K, Nicola A, Renteria T (2008) A proficiency testing method for detecting antibodies against Brucella abortus in quantitative and qualitative serological tests. Rev Sci Tech 27(3):819–828
4. Manning EJ, Collins MT (2001) Mycobacterium avium subsp. paratuberculosis: pathogen, pathogenesis and diagnosis. Rev Sci Tech 20(1):133–150
5. Nagamine K, Hase T, Notomi T (2002) Accelerated reaction by loop-mediated isothermal amplification using loop primers. Mol Cell Probes 16(3):223–229
6. Notomi T, Okayama H, Masubuchi H, Yonekawa T, Watanabe K, Amino N, Hase T (2000) Loop-mediated isothermal amplification of DNA. Nucleic Acids Res 28(12):E63
7. Trangoni MD, Gioffre AK, Ceron Cucchi ME, Caimi KC, Ruybal P, Zumarraga MJ, Cravero SL (2015) LAMP technology: rapid identification of Brucella and Mycobacterium avium subsp. paratuberculosis. Braz J Microbiol 46(2):619–626. doi:10.1590/S1517-838246220131206

Chapter 15

Highly Specific Ligation-dependent Microarray Detection of Single Nucleotide Polymorphisms

Noa Wolff and Ivan Barišić

Abstract

The fast detection and characterization of pathogens are essential for an efficient treatment of infectious diseases. However, the development of improved and reliable diagnostic methods is still an ongoing process because not only pathogens but also their antibiotic resistances have to be identified. The gold standard today is, however, a cultivation-based characterization approach, which takes days until results can be evaluated. In patients with, for example, severe sepsis, the diagnostic test duration is a very critical parameter because a delay of treatment optimization increases the mortality rate significantly. In contrast, DNA-based molecular techniques can obtain results within a few hours. A further challenge in diagnostic laboratories is that patient samples have to be screened for hundreds of potential pathogens, antibiotic resistance genes, and virulence factors, which is achieved by using a number of specialized tests at the moment. Microarrays are outstandingly good for the simultaneous analysis of thousands of different genes and have become a popular tool in biological studies. Nevertheless, further optimizations of the microarray technology are required due to the obligatory DNA labeling and/or amplification steps and the effects of nonspecific DNA hybridization. Here, we describe a fast and highly specific solid-support-based DNA characterization method for pathogens and antibiotic resistance genes.

Key words Molecular biology, Gene characterization, SNP, DNA microarray, Multiplex detection, Solid support-based detection

1 Introduction

The DNA microarray has emerged as a powerful tool in functional genome analysis and clinical diagnostics. It provides detailed information on the gene expression of an organism and allows the simultaneous detection of up to 100,000 target genes. The outstanding features of this high-throughput research technology have resulted in a wide range of applications and, as in our case, in the development of a species characterization microarray based on phylogenetic marker and antibiotic resistance genes including SNPs [1]. One of the limiting factors of this technique are the non-specific DNA hybridization events causing false-positive results [2, 3]. Various approaches to improve the sensitivity and

Kimberly A. Bishop-Lilly (ed.), *Diagnostic Bacteriology: Methods and Protocols*, Methods in Molecular Biology, vol. 1616, DOI 10.1007/978-1-4939-7037-7_15,

specificity of the microarray technology by using different spacers or surface chemistries, for example, were limited [4]. However, several experimental parameters could be identified that have an impact on the performance of microarrays. Specifically, the sensitivity and the specificity correlate with the length of the oligonucleotide probes [5]. Additionally, it was demonstrated by Chou et al. that the probe sensitivity is related to the length and the accessibility of the probe. Further effort to increase the specificity of the microarray technology resulted in the integration of a variety of enzymatic steps [6, 7].

Here, we demonstrate an improved multiplex detection and characterization method based on covalently immobilized DNA oligonucleotides combined with a ligation step. What is exceptional about our method is the short duration time and the specificity that is achieved by a novel type of immobilized probes, the linear chain (LNC) probes. This probe concept has already been used to identify clinically relevant pathogens [8]. The LNC probe is a thiol-modified detection probe that comprises three DNA oligonucleotides, LNC-A, LNC-B, and LNC-C (Fig. 1). The three oligonucleotides are connected via hydrogen bridge bonds. The GC-rich hybridization regions that link the three probes together have melting temperatures above 85 °C to facilitate a high LNC probe stability. Only LNC-C has a specific detection sequence for

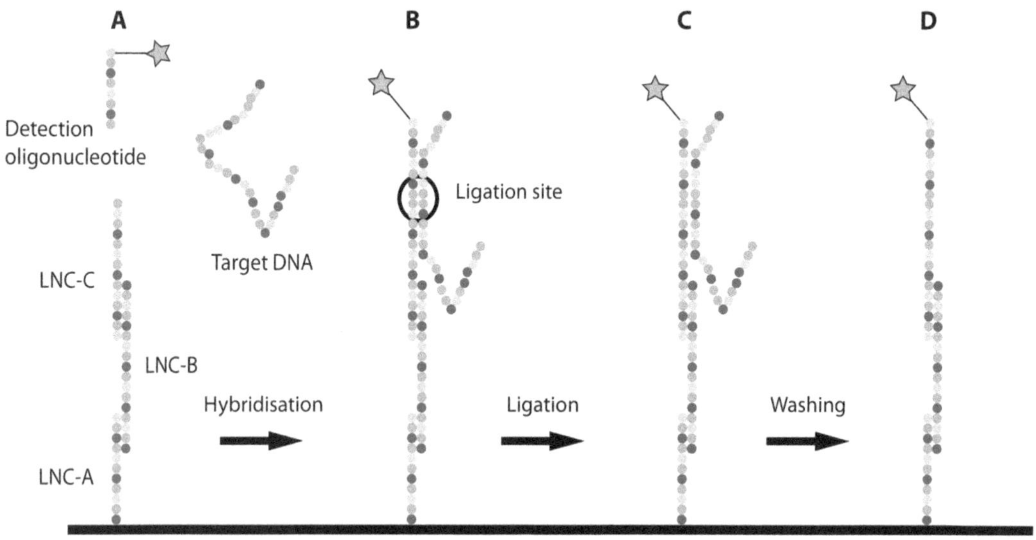

Fig. 1 Schematic illustration of the LNC probe and the reaction mechanism. (**a**) The LNC-A probe is linked covalently to a glass surface. LNC-A, B, and C are incubated together prior to spotting and immobilized in the hybridized state. A ligation mixture comprising fluorescently labeled detection oligonucleotides and the target DNA is applied to the slide. (**b**) The detection oligonucleotide binds in the presence of a target DNA to the LNC probe. (**c**) The ligation takes place if the target DNA is perfectly matching the LNC and the detection probe. (**d**) After the stringent washing step, only the ligated detection oligonucleotides remain on the surface while the non-ligated probes are washed away

the gene of interest. The LNC probes are immobilized to glass slides using a microarray spotter. In this protocol, we used bacterial pathogens and antibiotic resistance genes to illustrate the specificity of this method.

2 Materials

All buffers were prepared with deionized and micro-filtered water. The preparation and storage took place at room temperature. First, all glass trays and hybridization chambers were cleaned using DNA-Exitus (AppliChem, Germany) in order to be DNA free (*see* **Note 1**). The glass-slides were cleaned separately. The oligonucleotides were designed as follows:

LNC-A was modified at its 5′-end with a thiol group. The 3′-end of the detection oligonucleotides was modified with a Cy3-fluorophore. In addition, the 5′-end of the detection oligonucleotide has to be phosphorylated to ligate the oligonucleotide to the LNC-C probe. All oligonucleotides were purified by HPLC. The melting temperature for the target recognition region of the detection oligonucleotide and the LNC-C probe was about 50 °C. The sequences of the LNC oligonucleotides are shown in Table 1. The oligonucleotides were purchased from Integrated DNA Technologies (IDT, IA, USA).

2.1 Washing Buffer

1. For a 1 M HCl solution, add 83 ml 37% HCl in 600 ml $_{dd}H_2O$ and subsequently water to a total volume of 1 L (*see* **Note 2**).
2. For a 1 M NaOH solution, use 40 g NaOH and resolve them in 1 L water (*see* **Note 2**).

2.2 Silanization of Glass Slide

1. (3-Aminopropyl)trimethoxysilane ATS (Sigma-Aldrich, MO, USA): 0.5% solution in dry acetone (*see* **Note 3**). ATS forms an aminopropyl derivate on glass and is used as a first surface modifier.
2. Prepare PBS (0.1 M NaH_2PO_4, 0.15 M NaCl, pH 7.2).

Table 1
Structure and sequence of the LNC probes

Probe name	5′-Mod	Sequence 5′–3′	Length (bp)
LNC A	Thiol	TTTCGCTGCCGACCCTGCGCCGTGGCC	27
LNC B		CCCCGGCACGCGAGCCCACGCTGCTTTTTTGGCCACGGCGCAGGGTCGGCAGCG	54
LNC C		GCAGCGTGGGCTCGCGTGCCGGGGTTTTTTNNNNNNNNNNNNNNN	45

The sequence region with the multiple Ns indicates a variable region specific for the gene of interest. Ideally, it is approximately 15 base pairs (bp) long and has a melting temperature between 45 and 55 °C

3. Water-soluble heterobifunctional cross-linker: sulfonated analogs of *m*-maleimidobenzoyl-*N*-hydroxysulfosuccinimide ester (s-MBS) were purchased from ProteoChem (IL, USA) and prepared as 2 mM solution in PBS (*see* **Note 3**).

2.3 Spotting

1. 1× NaPi buffer: 0.1 M NaH_2PO_4, 0.15 M NaCl, pH 6.5 (*see* **Note 4**).
2. The oligonucleotides LNC-A, B, and C were pooled in a final concentration of 5 μM each in sterile filtered 1× NaPi buffer.
3. The LNC probes were spotted using the OmniGrid contact arrayer (GeneMachines, CA, USA) and SMP 3 pins (TeleChem, CA, USA).
4. β-Mercaptoethanol (Sigma-Aldrich): 10 mM β-mercaptoethanol in 1× NaPi-Buffer (*see* **Note 3**).
5. Saline buffer: 1.5 M NaCl, 10 mM NaH_2PO_4, pH 7.
6. 20× Saline sodium citrate (SSC) buffer, pH 7 (Invitrogen, CA, USA).
7. 5× SSC buffer comprising 0.1% Tween-20, pH 7.
8. 5× SSC buffer, pH 7.
9. Heraeus Megafuge 1.0 (Thermo Fisher Scientific) for slide centrifugation.

2.4 Ligation and Detection

1. First, prepare a stock mixture comprising all detection oligonucleotides with a final concentration of 100 μM. The end concentration of each oligonucleotide depends on the amount of the total number of oligonucleotides (*see* **Note 5**). Make sure that the 5′-ends of the detection oligonucleotides are phosphorylated. Oligonucleotides can be ordered either comprising the 5′-phosporylation or the modification can be introduced by the operator using, e.g., T4 polynucleotide kinase (Thermo Fisher Scientific) that is more economic in high-multiplex assays.
2. Prepare a master mixture containing the detection oligonucleotides with a final concentration of 300 nM for each detection oligonucleotide (*see* **Note 6**).
3. The ligation reaction was conducted in a frame-seal incubation chamber with a 25 μl capacity (Bio-Rad, CA, USA).
4. Bovine serum albumin (BSA, New England Biology, MA, USA): 2 μg/μl in $_{dd}H_2O$.
5. Ampligase buffer: 20 mmol/L Tris–HCl, 25 mmol/L KCl, 10 mmol/L $MgCl_2$, 0.5 mmol/L nicotinamide adenine dinucleotide (NAD) and 0.01% Triton® X-100, pH 8.3 (Epicentre, WI, USA).
6. 20% sodium dodecyl sulfate (SDS): 20 g SDS dissolved in 100 ml $_{dd}H_2O$ (*see* **Note 7**).

7. 2× SSC with 0.1% SDS washing buffer: 500 ml 20× SSC stock solution and 25 ml SDS 20% in 5 L $_{dd}H_2O$.
8. 0.2× SSC wash buffer: 50 ml 20× SSC stock solution in 5 L $_{dd}H_2O$.
9. The MJ Research PTC-200 Peltier Thermal cycler (Bio-Rad) was used for thermal incubation of the glass slides.
10. Slides were scanned using the Tecan PowerScanner (Tecan, Switzerland). Be aware that different microarray scanners have different limits of detection. Some scanners are not able to detect weak fluorescence signals. For analyzing the data, we used GenePix Pro 6.0 (Molecular Devices, CA, USA) and Excel 2007 (Microsoft, WA, USA).

3 Methods

Perform all steps at room temperature unless otherwise specified.

3.1 Cleaning of Glass Slides

1. Clean glass slides with $_{dd}H_2O$ followed by 100% EtOH (*see* **Note 8**).
2. Sonicate the slides for 10 min in acetone; subsequently wash twice with $_{dd}H_2O$ (*see* **Note 8**).
3. Afterward, sonicate the slides in 1 M NaOH for another 10 min and immerse in 1 M HCl overnight.
4. On the following day, wash the slides twice for 5 min in $_{dd}H_2O$; subsequently rinse with 100% EtOH and allow to air dry.

3.2 Silanization of Glass Slides

1. Immerse the dried and cleaned slides for 1 h in a 0.5% ATS solution in dry acetone (*see* **Notes 3** and **9**).
2. Afterward, wash the slides three times for 5 min with acetone and rinse them with 100% EtOH.
3. Subsequently, bake the slides for 50 min at 90 °C.
4. Wet the surface of the slides with 300 μl of the s-MBS solution in PBS buffer with a pH 7.4 for 5 h in a humid environment (*see* **Note 10**). The moist atmosphere can be obtained from an incubation chamber which is filled with water (*see* **Note 11**). This will counteract evaporation of your s-MBS solution.
5. While the cross-linker is incubated (*see* Fig. 2 for a schematic of the cross-linking reaction), the pipetting plate should be prepared for spotting, comprising NaPi buffer, the 5′-end thiol-modified oligonucleotide LNC-A, LNC-B and the target-specific LNC-C oligonucleotides (*see* **Note 12**). Dilute the LNC probes to a concentration of 5 μM each in 0.5× NaPi buffer. In addition, add two spotting controls to monitor the spotting efficiency and the LNC probe stability. The first

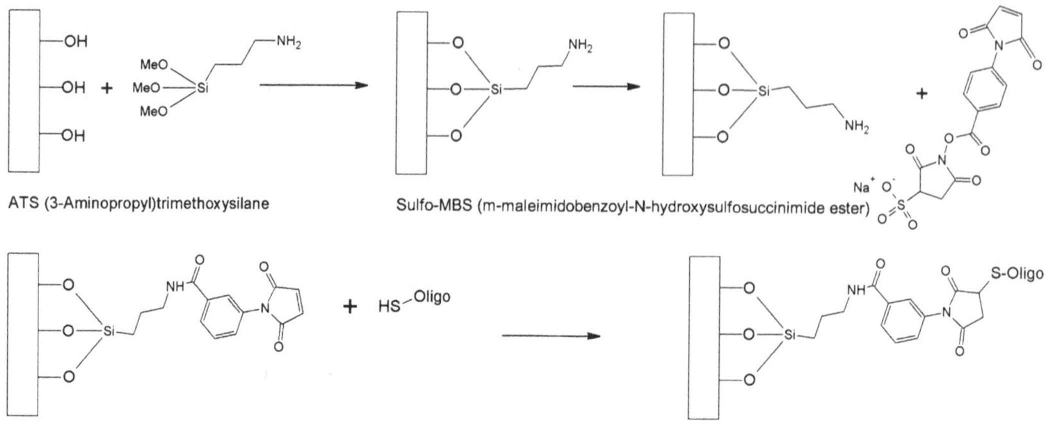

Fig. 2 Schematic illustration of the crosslinking reaction and the chemical attachment of the modified nucleic acid. This is a two-step reaction. First, the methoxy-group of the ATS reacts with the OH groups of the glass surface. Then, the s-MBS reacts with the amino group of ATS

control is a 5′ thiol-modified oligonucleotide with a fluorescently labeled 3′-end. The second control comprises the thiol-modified LNC-A and a fluorescently labeled LNC-B oligonucleotide hybridized to each other.

6. After the incubation with the s-MBS solution, rinse the slides with PBS.
7. Remove excess salts by washing twice with $_{dd}H_2O$.
8. Immerse the slides in 100% EtOH and allow to air dry overnight.
9. Spot the LNC probes on the glass slides by using the OmniGrid contact arrayer at an adjusted air humidity of 60% (*see* **Note 13**). Incubate the oligonucleotides for 5 h in a humid incubation chamber. The subsequent washing steps should be conducted in the dark.
10. Wash the slides in 1× NaPi buffer for 5 min.
11. To deactivate the reactive groups on the surface, incubate the slides with 10 mM β-mercaptoethanol in 1× NaPi for 1 h (*see* **Note 14**).
12. Afterward, wash the slides again in 1× NaPi buffer for 5 min.
13. Precipitate unbound oligonucleotides in saline buffer for 10 min.
14. Then, wash the slides in 5× SSC buffer comprising 0.1% Tween for 5 min.
15. Wash the slides in 5× SSC buffer for 1 min.
16. Finally, wash the slides in$_{dd}H_2O$ twice and centrifuge them to dry (*see* **Note 15**).
17. Store the slides at −20 °C.

3.3 Ligation and Detection

1. After warming the slides to room temperature, apply the frame-seal incubation chambers to the slides (*see* **Note 16**).
2. Prepare the reaction mixture comprising the ampligase buffer, 6 μg of BSA, 7.5 U of ampligase and detection oligonucleotides (300 nM each) in a total volume of 30 μl (*see* **Note 17**). One microliter of target DNA must be added separately to each individual reaction.
3. Pipette the reaction mixture into the reaction chamber and cover the chamber with the provided foils.
4. Perform the ligation in the slide cycler for 1 h at 55 °C.
5. After the ligation, wash the slides with 2× SSC buffer (0.1% SDS) for 5 min.
6. Subsequently, wash the slides in 0.2× SSC buffer for 2 min.
7. Finally, wash the slides in $_{dd}H_2O$ for 1 min (*see* **Note 18**).

   Optional: To efficiently differentiate SNPs, a stringent washing step in $_{dd}H_2O$ at 70 °C for 5–10 min has to be conducted. The non-ligated detection oligonucleotides that hybridize via mismatching target DNA to the LNC probe are removed in this step. In contrast, the detection oligonucleotides that are ligated to the LNC C probe are covalently attached to the LNC probe and can withstand stringent washing steps of 70 °C and higher (Fig. 3).
8. Dry the slides by centrifugation for 1 min at 900 rpm (*see* **Note 19**).
9. Scan the slides with a microarray scanner and analyze.

4 Notes

1. If you are using DNA-Exitus make sure you remove it well with plenty of water. It will not only destroy your probes and enzymes but also change the pH-value of your buffers.
2. Having water first in the flask prevents super heating.
3. This solution should be prepared each time afresh. s-MBS, such as most cross-linkers, is moisture-sensitive. Additionally, avoid contamination with primary amines that compete with your actual reactant, the amine of the ATS group.
4. pH 6.5 is necessary to limit disulfide bond formation between thiol-modified oligonucleotides (LNC-A probes).
5. Fluorophores are light sensitive; handle them in the dark. If you are using a PCR hood with UV light, be aware that ozone generated by UV light destroys fluorescence molecules as well.
6. Prepare these steps in a clean PCR hood to avoid DNA contaminations.

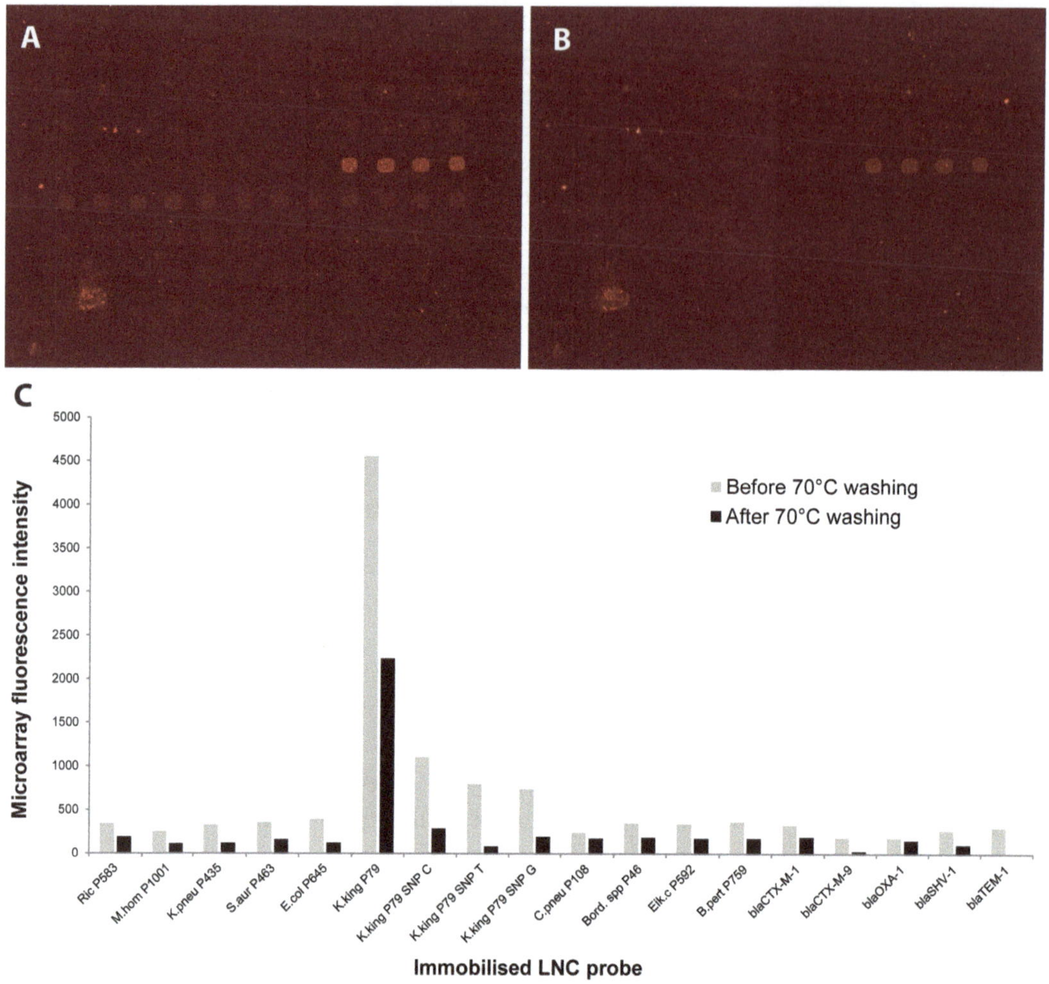

Fig. 3 Microarray results from a 25-multiplex experiment illustrating the SNP detection specificity [8]. Microarray images of the slides (**a**) before the stringent washing step and (**b**) after the stringent washing step. The brightness and the contrast values were set to the same levels in both images. (**c**) Chart showing the microarray fluorescence intensities before and after the washing step

7. SDS is hard to dissolve but preheating the water (37 °C) facilitates the process together with a stir bar (Caution: SDS powder is hazardous). Weigh and prepare the solution in a ventilated fume hood or use a dust mask. SDS precipitates at temperatures below 15 °C.
8. Make sure that you carefully wash the slides. They have to be DNA free to avoid contaminations.
9. Make sure that the slides and the reaction chamber in which you process the ATS incubation are water free. Water molecules inhibit the modification of your glass surfaces. You may centrifuge the slides until they are dry or by blowing off the liquid

before modifying the surface. However, air drying is the best solution in order to avoid DNA contaminations.

10. We use a sandwich setup with two slides for the incubation with s-MBS. We apply 300 μl of the s-MBS solution to one slide and put a second slide on top of it. To provide a reaction space between the slides, we use parafilm stripes as spacers on the edges of the slides. If you apply the s-MBS solution to the slides, take care that no foam is pipetted. When adding the top slide, take care that no air bubbles enter the inter-slide area.
11. If you have to move the humidity chamber make sure that the water that you use to create a humid atmosphere is not splashing on your slides. To avoid this, use tissue; it absorbs the water but still releases moisture.
12. Use a 396-well spotting plate. Conduct the liquid manipulation steps in a PCR hood to prevent contaminations. You can use this plate multiple times. Make sure to avoid evaporation.
13. Use one dummy slide for spotting. This serves to highlight the spotting area on your slides and will help you to stick the reaction chamber properly at your slides.
14. The glass slides were incubated in β-mercaptoethanol to cap residual maleimide active moieties. β-Mercaptoethanol is toxic and smelly; therefore, we strongly recommend using it only under the hood.
15. If you centrifuge your glass slides do not use a speed above 900 g. Higher speed may destroy your slides and your centrifuge.
16. Pay close attention how you stick the reaction chambers to the slides. By misapplication you can destroy the spots.
17. Fluorophores are very light sensitive. They have to be handled in the dark. Also be aware that ozone also destroys them. PCR hoods with an UV-light filter generate ozone that is still present even after switching off the UV-light in workplace area.
18. Leftover salts can increase the background fluorescence signals of the slides. An additional washing step in $_{dd}H_2O$ can be introduced if problems with high background fluorescence occur.
19. Be aware that water can destroy your scan equipment. Make sure that your slides are dry when you analyze them with a microarray scanner.

Acknowledgment

We thank the technical staff at the AIT for providing technical assistance and Christine Giuffrida for the proofreading of the manuscript.

References

1. Barišic I, Petzka J, Schoenthaler S et al (2015) Multiplex characterisation of human pathogens including species and antibiotic resistance gene identification. J Med Microbiol. doi:10.1099/jmm.0.000192
2. Wilkes T, Laux H, Foy CA (2007) Microarray data quality—review of current developments. OMICS 11(1):1–13
3. Dai H, Meyer M, Stepaniants S et al (2002) Use of hybridization kinetics for differentiating specific from non-specific binding to oligonucleotide microarrays. Nucleic Acids Res 30(16): e86
4. Koltai H, Weingarten-Baror C (2008) Specificity of DNA microarray hybridization: characterization, effectors and approaches for data correction. Nucleic Acids Res 36(7):2395–2405
5. Chou C, Chen C, Lee T et al (2004) Optimization of probe length and the number of probes per gene for optimal microarray analysis of gene expression. Nucleic Acids Res 32(12):e99. doi:10.1093/nar/gnh099
6. Ericsson O, Jarvius J, Schallmeiner E et al (2008) A dual-tag microarray platform for high-performance nucleic acid and protein analyses. Nucleic Acids Res 36(8):e45
7. Adessi C, Matton G, Ayala G et al (2000) Solid phase DNA amplification: characterisation of primer attachment and amplification mechanisms. Nucleic Acids Res 28(20):E87
8. Barišić I, Kamleithner V, Schönthaler S et al (2014) Fast and highly specific DNA-based multiplex detection on a solid support. Appl Microbiol Biotechnol. doi:10.1007/s00253-014-6246-x

Chapter 16

Multilocus Sequence Typing (MLST) for *Cronobacter* spp.

Susan Joseph and Stephen Forsythe

Abstract

MLST is a molecular typing technique that involves the identification and clustering of bacterial isolates based on the partial sequence analysis of multiple housekeeping genes (generally seven) which are distributed across the length of the genome of the organism. The *Cronobacter* whole genus MLST scheme can be successfully used for an accurate species level identification and classification of this complex genus.

Key words *Cronobacter*, MLST, Sequence type, Clonal complex, PubMLST, Neonatal meningitis, ST4

1 Introduction

Cronobacter (previously known as *Enterobacter sakazakii*) is a Gram-negative bacterial genus notable for causing illnesses such as neonatal meningitis and necrotising enterocolitis in both infants and adults [1]. The genus, having been reclassified from the *E. sakazakii* species, now consists of seven formally recognized species, namely, *C. sakazakii*, *C. malonaticus*, *C. turicensis*, *C. muytjensii*, *C. dublinensis*, and the more recently characterized *C. universalis* and *C. condimenti* [2, 3].

The identity of the organism is further complicated by the fact that both 16S rDNA sequence analysis and phenotypic biotyping tests fail to accurately provide species level characterization [4]. This problem was overcome with the establishment of the 7-loci Multi Locus Sequence typing (MLST) scheme, which was initially established to distinguish between the species *C. sakazakii* and *C. malonaticus*, and later expanded to cover the entire genus [5, 6].

The *Cronobacter* MLST scheme consists of the seven genes, a*tpD*, *fusA*, *glnS*, *gltB*, *gyrB*, *infB*, and *ppsA*. All details of alleles, sequence types (STs), clonal complexes (CCs), as well as isolates are stored in an open-access, curated database at http://pubmlst.org/cronobacter/. The database currently houses details of 1568 bacterial strains, which comprise 466 STs and 53 CCs, of which

Kimberly A. Bishop-Lilly (ed.), *Diagnostic Bacteriology: Methods and Protocols*, Methods in Molecular Biology, vol. 1616,
DOI 10.1007/978-1-4939-7037-7_16,

CC4 has now been established as a genetic signature for neonatal meningitis infections caused by *Cronobacter* [1, 7, 8].

In recent years, the database has also been upgraded to the Bacterial Isolate Genome Sequence Database (BIGSdb) platform, which enables users to perform not just the traditional 7-loci MLST but also involves the molecular characterization of strains at the whole genome level using other associated MLST schemes such as the ribosomal MLST (rMLST; 51 loci), Clusters of Orthologous Genes (COG)-cgMLST (1865 loci), and the Tax-MLST (*ompA* & *rpoB* for taxonomic evaluations) [9].

Here, we present a step-wise protocol to perform the conventional 7-loci MLST for the *Cronobacter* genus, using a combination of wet laboratory techniques and analysis tools from the MLST database.

2 Materials

2.1 PCR Amplification of the Seven MLST Genes

1. Approximately 10 ng of DNA is needed for each PCR reaction to obtain a sufficient amount of amplified template for sequencing (*see* **Note 1**). Only DNA samples with minimum (260/280) nm values of 1.8 and (260/230) nm values of 2 are to be used, else DNA extraction must be repeated.
2. Primers are to be used from a working stock concentration of 10 μM each, dilutions to be made by using molecular biology grade dH_2O. Primer sequences for PCR have been listed in Table 1.

Table 1
Primer sequences for the seven loci used in the *Cronobacter* MLST scheme

Target	Putative function	PCR primer sequence (5′–3′)	Expected product size (bp)
atpD	ATP synthase β chain	CGACATGAAAGGCGACAT TTAAAGCCACGGATGGTG	998
fusA	Elongation factor	GAAACCGTATGGCGTCAG AGAACCGAAGTGCAGACG	1376
glnS	Glutaminyl tRNA-synthetase	GCATCTACCCGATGTACG TTGGCACGCTGAACAGAC	824
gltB	Glutamate synthase large subunit	CATCTCGACCATCGCTTC CAGCACTTCCACCAGCTC	2091
gyrB	DNA gyrase β subunit	TGCACCACATGGTATTCG CACCGGTCACAAACTCGT	1946
infB	Translation initiation factor IF-2	GAAGAAGCGGTAATGAGC CGATACCACATTCCATGC	1470
ppsA	Phosphoenol pyruvate synthase	GTCCAACAATGGCTCGTC CAGACTCAGCCAGGTTTG	2358

3. The following components are needed to set up the PCR master mix:
   - 5× Green GoTaq Buffer (Promega).
   - 25 mM $MgCl_2$ solution (Promega).
   - dNTP mix, 10 mM each (Promega).
   - GoTaq DNA Polymerase, 5 U/μl (Promega).
4. Depending on the number of samples, PCR may be carried out in 0.2 ml PCR tubes or in 96-well plates.

2.2 Agarose Gel Electrophoresis

1. Running buffer: Make 1000 ml of 1× Tris-Acetate-EDTA (TAE) buffer by adding 20 ml of commercial 50× TAE buffer to 980 ml dH_2O.
2. 1% Agarose gel: Weigh 1 g of agarose and add it to 100 ml of 1× TAE buffer in a conical flask. Heat the solution in a microwave to completely dissolve the agarose. Add 5 μl of SYBR® Safe DNA gel stain (Life Technologies, Invitrogen, UK) and stir well to dissolve uniformly. Pour in a suitable sized electrophoresis tray with combs and let the gel set for 30–45 min.

2.3 PCR Purification

1. MinElute PCR Purification Kit (Qiagen).
2. Molecular biology grade water.

3 Methods

3.1 Polymerase Chain Reaction

1. The volumes of the master mix components for each 25 μl PCR reaction are to be aliquoted, as described in Table 2. Scale up the volumes according to the number of samples to

Table 2
PCR master-mix reaction volumes for the *Cronobacter* MLST scheme

	1×
5× Green GoTaq Buffer	5 μl
25 mM $MgCl_2$ solution	1.5 μl
dNTP mix, 10 mM each	2 μl
Forward Primer, 10 μM	2 μl
Reverse Primer, 10 μM	2 μl
GoTaq DNA Polymerase, 5 U/μl	0.25 μl
DNA	1 μl
Mol. Biology grade dH_2O	11.25 μl
Total	25 μl

be amplified, also accounting for a 10% pipetting error. Also include a no-template reaction to be used as a negative control for each set of primers.

2. Place the reaction tubes in the thermal cycler and run the program with the following conditions: initial denaturation at 94 °C for 2 min; 30 cycles of denaturation at 94 °C for 1 min, primer annealing at 58 °C for 1 min, extension at 72 °C for 2 min; followed by a final extension step of 72 °C for 5 min.

3.2 Agarose Gel Electrophoresis

1. Load 5 μl aliquots of each PCR reaction into each well of the agarose gel.
2. Load the first or the last well with 5 μl of 1 kb DNA ladder (Promega, UK) to be used as a marker for size comparisons.
3. Electrophoresis settings: 100 V for 40 min in 1× TAE running buffer.
4. View the gel under ultraviolet light to observe the amplified DNA bands. Expected band sizes for each gene are also listed in Table 1.

3.3 PCR Purification and Sequencing

1. Once it has been confirmed that only a single band of expected size has been amplified, use the remaining 20 μl of the PCR reaction to purify the product using the Qiagen MinElute PCR purification kit according to manufacturer's instructions.
2. At the final step of the purification protocol, elute the purified DNA product in 50 μl of molecular biology grade water.
3. Quantify the DNA and check the quality using a Nanodrop and then dilute the DNA to a final concentration of 10 ng/μl.
4. Use the diluted DNA sample for sequencing the amplified gene using the primers listed in Table 3. Using these nested sequencing primers, the nucleotide sequence will be determined at least once on each DNA strand.

3.4 Sequence Analysis

1. Check the quality of the ABI format output sequence files for each gene using a program such as ChromasLite (Technelysium Pty Ltd) to ensure that the bases have been accurately called (*see* Notes).
2. Align the forward and reverse sequences using the multiple sequence alignment tool MUSCLE (http://www.ebi.ac.uk/Tools/msa/muscle/) [10] to obtain a consensus region. Export this region into a basic text editor and save as a FASTA sequence file (*.fas/*.fasta). Do this for each of the seven genes of the MLST scheme.

3.5 Allele and Sequence Type Designation

1. The allele designation for each of these aligned sequences can be obtained at the *Cronobacter* PubMLST database at http://pubmlst.org/cronobacter/.

Table 3
Sequencing primers and target allele lengths for the 7-loci *Cronobacter* MLST

Target	Sequencing primer sequence (5′–3′)	Target allele size (bp)
atpD	CGAAATGACCGACTCCAA GGATGGCGATGATGTCTT	390
fusA	GCTGGATGCGGTAATTGA CCCATACCAGCGATGATG	438
glnS	GGGTGCTGGATAACATCA CTTGTTGGCTTCTTCACG	363
gltB	GCGAATACCACGCCTACA GCGTATTTCACGGAGGAG	507
gyrB	CTCGCGGGTCACTGTAAA ACGCCGATACCGTCTTTT	402
infB	TGACCACGGTAAAACCTC GGACCACGACCTTTATCC	441
ppsA	ACCCTGACGAATTCTACG CAGATCCGGCATGGTATC	495

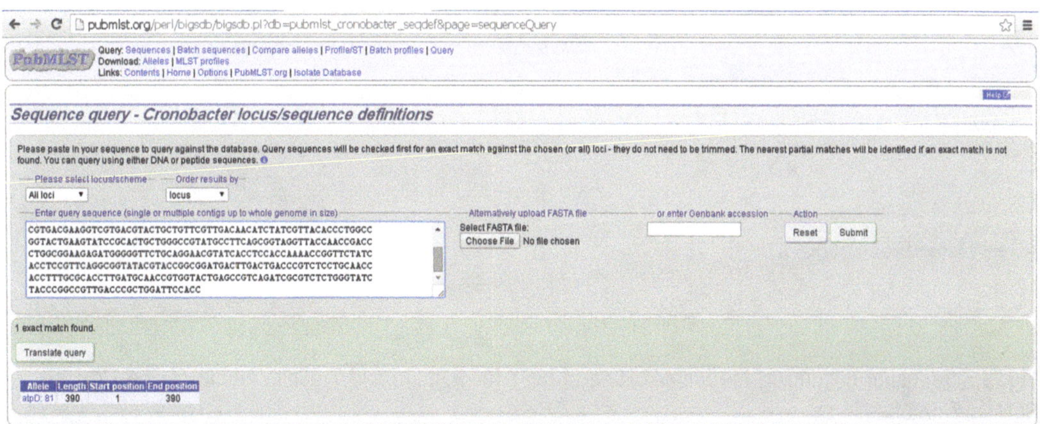

Fig. 1 Screenshot describing the submission of an allele sequence query on the *Cronobacter* PubMLST database

2. From the homepage, follow the tabs: Sequence and profile definitions > Query database > Sequence query

   Submit the sequence to be assigned either by pasting the sequencing or by uploading the FASTA file and the database will output a result that indicates the allele number for the sequence (Fig. 1).

3. If the sequence has been previously unidentified, the output will indicate that and the sequence will have to be submitted to the curator of the database to be assigned an allele number.

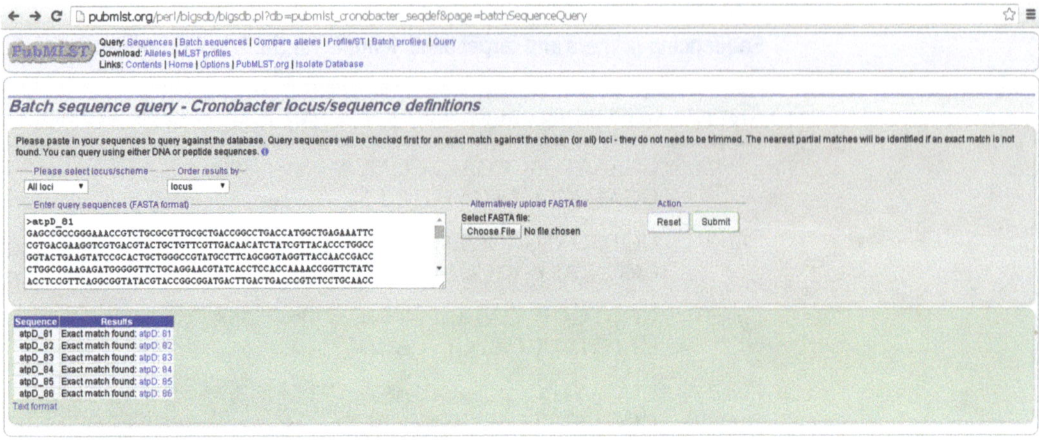

Fig. 2 Screenshot describing the submission and output of a batch sequence query on the *Cronobacter* PubMLST database

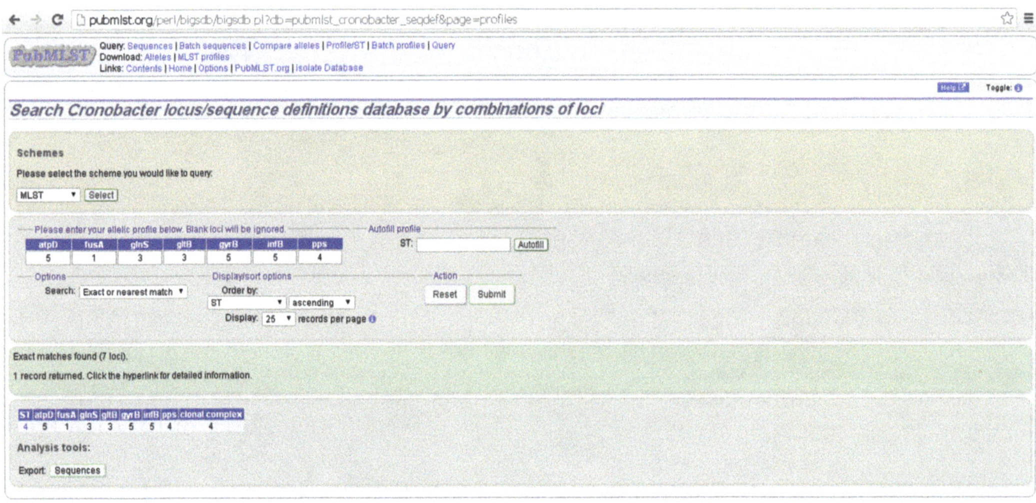

Fig. 3 Screenshot describing the identification of a ST for a submitted 7-allelic profile on the *Cronobacter* PubMLST database

4. Multiple sequences can be analyzed together by following the "Batch sequence query" under the Query database tab (Fig. 2).
5. Once all seven genes of a bacterial strain have allele numbers assigned to them, go to Sequence and profile definitions > Query database > Search by combinations of alleles. Select the "MLST" scheme from the drop-down menu and key in the allele numbers for each of the seven genes of the strain. When the query is submitted, the database outputs the result indicating the sequence type (ST) corresponding to the submitted combination of alleles (Fig. 3). As earlier, if the submitted combination of alleles is of a previously unidentified ST, then the database curator will assign it a new and unique ST number.

3.6 Submission of Newly Identified Alleles and Sequence Types

1. Researchers can submit any newly identified alleles or sequence types as well as details of typed *Cronobacter* strains for inclusion in the database.
2. All the required information needs to be emailed to the curator of the database, Prof. Stephen Forsythe (sforsythe4j@gmail.com) who will quality check the data before uploading it to the database. More details and templates for submission are available at the following link: http://pubmlst.org/cronobacter/submission.shtml.
3. It is necessary to submit the trace sequencing files along with the sequences of the newly identified alleles, in order to facilitate the curating process.

4 Notes

1. Even though a 10 ng starting material is recommended, samples with low yield DNA may be attempted. Care must be taken, however, to ensure that adequate product is obtained for successful sequencing. This may be done by scaling up the PCR reaction volumes to 50 μl and eluting the purified product to a reduced volume of ~30 μl to ensure adequate concentrations.
2. Careful perusal of the sequence chromatogram is very important to ensure accurate results. It is important to check for any observed background contamination or miscalled nucleotide bases as this can influence the outcome of the allelic profile and thereby the MLST profile.
3. Additional help with analyzing the sequences in the *Cronobacter* PubMLST database for yourself can be obtained by watching the YouTube and YouKU video guides on the homepage (http://pubmlst.org/cronobacter/).

References

1. Joseph S, Forsythe S (2011) Predominance of *Cronobacter sakazakii* sequence type 4 in neonatal infections. Emerg Infect Dis 17: 1713–1715
2. Iversen C, Mullane N, McCardell B et al (2008) *Cronobacter* gen. nov., a new genus to accommodate the biogroups of *Enterobacter sakazakii*, and proposal of *Cronobacter sakazakii* gen. nov.,comb. nov., *Cronobacter malonaticus* sp. nov., *Cronobacter turicensis* sp. nov., *Cronobacter muytjensii* sp. nov., *Cronobacter dublinensis* sp. nov., *Cronobacter* genomospecies 1, and of three subspecies, *Cronobacter dublinensis* subsp. dublinensis subsp. nov., *Cronobacter dublinensis* subsp. lausannensis subsp. nov. and *Cronobacter dublinensis* subsp. lactaridi subsp. nov. Int J Syst Evol Microbiol 58:1442–1447
3. Joseph S, Cetinkaya E, Drahovska H et al (2012) *Cronobacter condimenti* sp. nov., isolated from spiced meat, and *Cronobacter universalis* sp. nov., a species designation for *Cronobacter* sp. genomospecies 1, recovered from a leg infection, water and food ingredients. Int J Syst Evol Microbiol 62:1277–1283
4. Jackson EE, Forsythe SJ (2016) Comparative study of *Cronobacter* identification according to phenotyping methods. BMC Microbiol 16:146

5. Baldwin A, Loughlin M, Caubilla-Barron J et al (2009) Multilocus sequence typing of *Cronobacter sakazakii* and *Cronobacter malonaticus* reveals stable clonal structures with clinical significance which do not correlate with biotypes. BMC Microbiol 9:223
6. Joseph S, Sonbol H, Hariri S et al (2012) Diversity of the *Cronobacter* genus as revealed by multilocus sequence typing. J Clin Microbiol 50:3031–3039
7. Hariri S, Joseph S, Forsythe SJ (2013) *Cronobacter sakazakii* ST4 strains and neonatal meningitis, US. Emerg Infect Dis 19:175–177
8. Joseph S, Forsythe SJ (2012) Insights into the emergent bacterial pathogen *Cronobacter* spp., generated by multilocus sequence typing and analysis. Front Food Microbiol 23:397
9. Forsythe SJ, Dickins B, Jolley KA (2014) Cronobacter, the emergent bacterial pathogen *Enterobacter sakazakii* comes of age; MLST and whole genome sequence analysis. BMC Genomics 15:1121
10. Goujon M, McWilliam H, Li W et al (2010) A new bioinformatics analysis tools framework at EMBL–EBI. Nucleic Acids Res 38(Suppl 2): W695–W699

Chapter 17

Diagnostic Bacteriology: Raman Spectroscopy

Rebecca L. Pavlicek, Nicole J. Crane, Meron Ghebremedhin, Katherine E. Cilwa, and Eric A. Elster

Abstract

Current clinical methodology for identification of bacterial infections relies predominantly on culturing microbes from patient material and performing biochemical tests. This can often be an inefficient and lengthy process, which has a significant detrimental effect upon patient care. Techniques used in other aspects of molecular research have the potential to revolutionize the way in which diagnostic tests are used and delivered in the clinical setting. The need for rapid, accurate, and cost-effective molecular techniques in the diagnostic laboratory is imperative to improving patient care, preventing the spread of drug resistance and decreasing the overall burden associated with nosocomial infections. Raman spectroscopy and surface-enhanced Raman spectroscopy (SERS) are powerful vibrational spectroscopy techniques that are being developed for highly sensitive pathogen identification in complex clinical samples. Raman spectroscopy is a molecular technique that is capable of probing samples noninvasively and nondestructively. It has been used with high specificity to assess tissue and bacterial samples at the molecular level with diverse clinical and diagnostic applications. SERS has recently developed out of the advances in the Raman spectroscopy arena. This technique is designed to amplify Raman scattering and allows for better differentiation of bacterial isolates. Although the current parameters for the use of SERS require a pure culture and are relatively monoparametric, current breakthroughs and testing are pushing the technology to new levels and thus changing the face of modern bacterial diagnostics.

Key words Molecular diagnostics, Bacteriology, Clinical microbiology, Rapid diagnostic, Bacteria, Infection, Antimicrobial resistance, Bacterial profiling, Raman spectroscopy, Surface enhanced, SERS

1 Introduction

Although molecular diagnostics have continued to evolve rapidly, the protocols in many clinical bacteriology laboratories remain practically unchanged from those used 75 years ago [1]. Even with a tangible need in the clinic, this field has remained stubbornly resistant to the adaptation and utilization of tools from other scientific fields. The diagnostic bacteriology field has been broken into two major components: pathogen identification and resistance profiling [1, 2]. While these areas are considered separate components within the field, they often go hand-in-hand and have

Kimberly A. Bishop-Lilly (ed.), *Diagnostic Bacteriology: Methods and Protocols*, Methods in Molecular Biology, vol. 1616, DOI 10.1007/978-1-4939-7037-7_17, © Springer Science+Business Media LLC 2017

a direct effect upon each other. Even though the field has seen a dramatic change in definitive diagnosis since the advent of sequencing and other molecular techniques in the early 1980s, one of the most significant challenges for the diagnostic laboratory continues to be the time required for organism identification and typing [1]. To date, this area of medical study has lagged behind and current techniques remain inadequate. It often takes days to weeks for laboratories to isolate, grow, and identify pathogens, much less to provide a detailed analysis of the bacterium and its resistance profile. In order to provide the best patient care and remain cognizant of antimicrobial stewardship, physicians require rapid as well as accurate diagnostics. This time delay has led to an overuse of antimicrobials and "shot-gun" treatments in the clinical setting. Many patients are unable to handle any delay in treatment, so physicians are required to start broad spectrum treatments until a definitive diagnosis has been reached. Additionally, in critical care settings, rapid identification is the key to preventing the spread of nosocomial pathogens to other patients as well as employees. Techniques used in other aspects of molecular research have the potential to revolutionize the way in which diagnostic tests are used and delivered in the clinical setting. The need for rapid, accurate, and cost-effective molecular techniques in the diagnostic laboratory is imperative to improving patient care, preventing the spread of drug resistance and decreasing the overall burden associated with nosocomial infections.

Since the discovery of sequencing and other molecular techniques, an enormous amount of research has enabled the introduction and routine use of molecular tests in several areas of medical research while bacterial diagnostic laboratories have been slow to change methodology. Many laboratories are reluctant to introduce new technology and methods due, in part, to concerns over cost, reliability, and training requirements. As a result, a large number of hospital-based diagnostic laboratories do not routinely use any molecular diagnostic techniques for bacterial identification and resistance profiling, continuing to use traditional culturing methods. Although the need for change has chipped away at this reluctance to integrate technology, there remains a significant gap. As the use of molecular techniques becomes more widespread, the requirement and expectation of a basic understanding of these techniques will be necessary for many medical professionals. This chapter is intended to explore two of those techniques, Raman spectroscopy and Surface Enhanced Raman Spectroscopy (SERS), in the diagnostic bacteriology setting.

1.1 Raman Spectroscopy and Surface-Enhanced Raman Spectroscopy (SERS)

While Raman spectroscopy is widely used in the chemistry field, flow cytometry and fluorescence microscopy remain the methods of choice in the molecular biology arena. Raman spectroscopy is a molecularly specific technique that is capable of probing samples noninvasively and nondestructively. It has been used with high

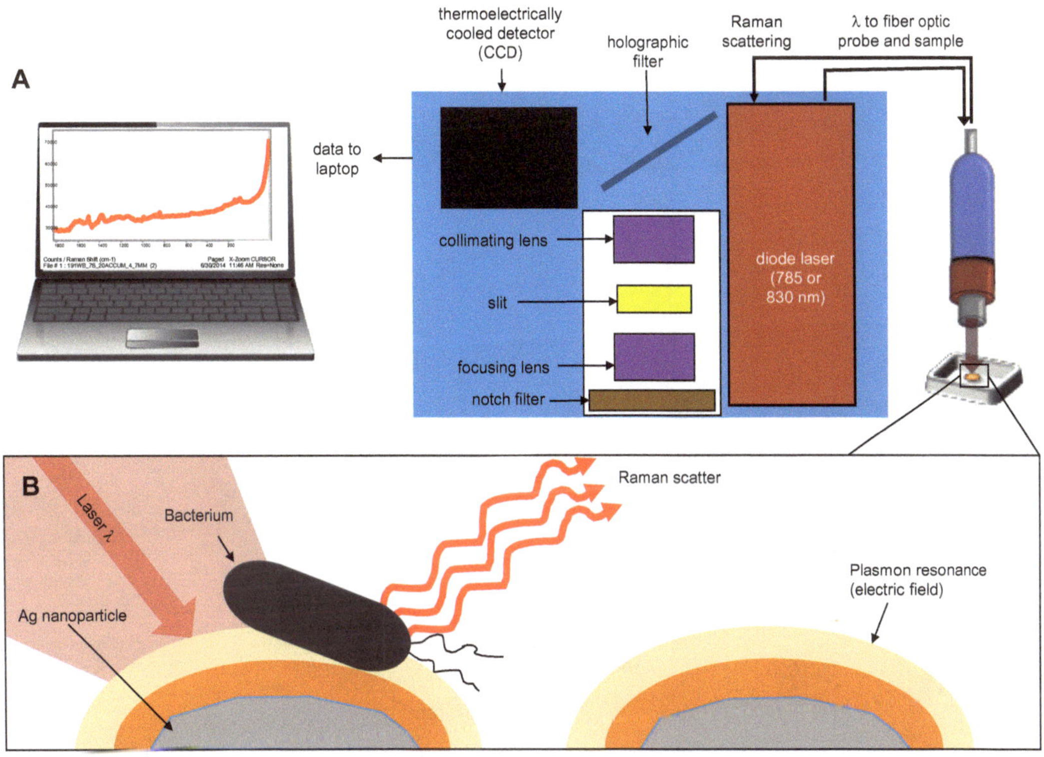

Fig. 1 Diagram of setup and internal components of SERS equipment (**a**) and SERS enhancement (**b**)

specificity to assess tissue and bacterial samples at the molecular level with diverse clinical and diagnostic applications with high specificity [3–19]. This is one of the attributes that made Raman spectroscopy an ideal technology for further development in the field of bacterial diagnostics.

Raman spectroscopy employs a laser that, when focused on a sample, produces an identifiable spectrum that provides information about the sample based on its molecular vibrations (Fig. 1) [20]. Although the intensity of Raman scattering is incredibly less efficient than fluorescence, some of the issues associated with this technique have been mitigated by the use of plasmonic nanoparticles and other nanostructures to achieve surface-enhanced Raman spectroscopy (SERS) [21]. These nanostructures, when excited with an appropriate wavelength, increase the intensity of Raman scattering to the levels measured with other techniques such as fluorescence. Signal enhancement up to 10^{10}–10^{11} has been observed, allowing single molecule detection in certain systems (Fig. 2).

SERS enhancement is controlled by three main factors. First, enhancement is initiated by the excitation of plasmonic states on the surface of SERS-active substrate upon interaction of the

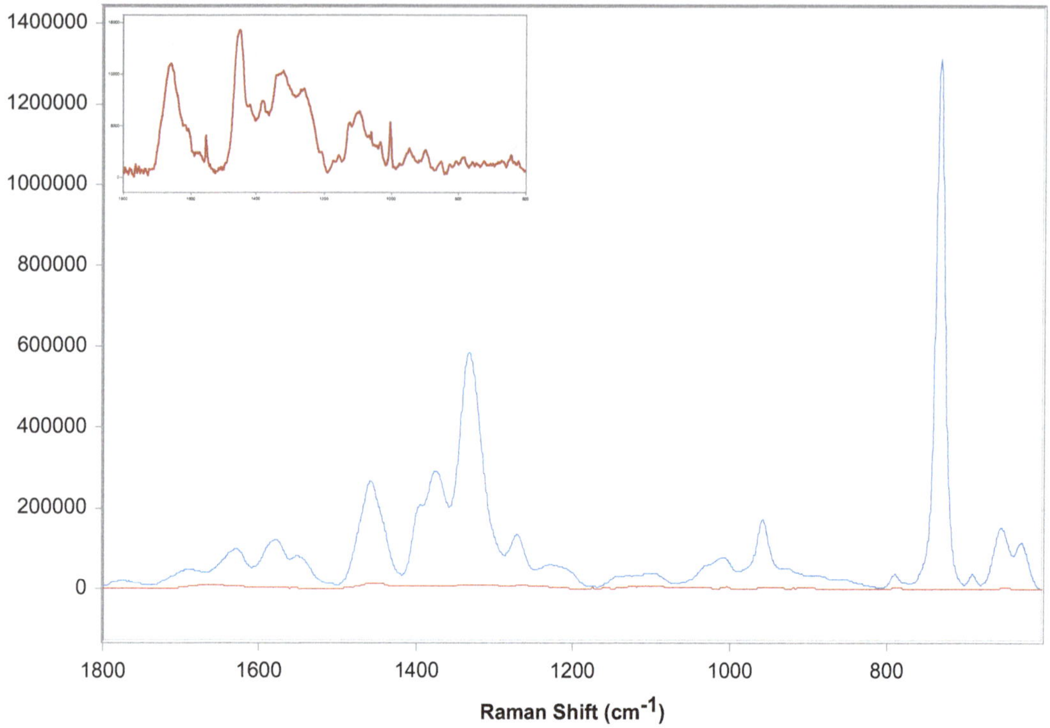

Fig. 2 SERS enhancement is demonstrated below with a Raman spectrum (*red line*) and an overlaid SERS spectrum (*blue line*). The unenhanced Raman spectrum is shown in the inset on a smaller scale to magnify the vibrational bands

substrate with incident light. This results in concentrated, enhanced electric fields at the surface of the substrate. Raman scattering of analytes within the proximity of these enhanced electric fields may greatly increase. Enhancement may be achieved when analytes are either chemically bonded to the substrate or physically adsorbed [21]. This factor plays the most important role in signal boost. Second, broad fluorescence signal often dominates the collected signal in Raman spectroscopy. Fluorescence may be greatly reduced by inherent quenching mechanisms of SERS-active substrates. Lastly, further enhancement and specificity of SERS spectra are achieved by reduced band overlap of Raman active peaks and addition of peaks usually only captured using infrared spectroscopy to the observed spectrum yielding more unique variance in the data from species to species due to analyte-substrate interactions.

SERS-active substrates may be composed of monomers, dimers, and arrays of nanoparticles as well as other nanostructured metal surfaces. Systems are highly customizable based on the size, shape, and material employed. Substrates may be functionalized to selectively detect and/or capture molecules, cells, or proteins of interest using reporter molecules or antibodies. The synthesis of unique structures such as nanoshells, porous nanoparticles, and layered substrates results in unique systems that can utilize magnetic properties to capture analytes. The addition of SERS-active substrates to microfluidic devices allows for detection at low concentration [22]. The first observation of SERS occurred over 40 years ago and initial research focused largely on mechanistic studies. Since then SERS has been utilized in a wide variety of applications including in the detection of cancer cells and biomarkers of interest. Nanoparticle-based SERS substrates may be used to monitor molecules both inside and outside the cell. Focus on the use of SERS for bacterial detection, identification, and monitoring has rapidly increased in the last decade.

1.2 SERS and Raman Spectroscopy in the Clinical Setting

The application of spectroscopic methods, particularly SERS and Raman spectroscopy, in a clinical setting has yet to be fully explored and integrated. There have been numerous Raman spectroscopic studies of microorganisms, many focusing on rapid identification of the bacterial isolates [23–31].

Current methods of identifying bacterial infection rely on culturing microbes from patient material and performing biochemical tests, which together can take 2–3 days to complete. If SERS could detect bacterial infection from patient material directly, physicians would be able to determine course of treatment and drug administration in a matter of hours rather than days. Its potential benefits could revolutionize the speed and reliability of pathogen identification in the clinic as well as other venues. This could greatly affect wound outcomes, particularly in combat-related injuries that tend to be significantly traumatic and multivariable, which has been the focus of our efforts.

In previous spectroscopic studies, bacterial isolates were successfully identified at the strain level by utilizing a Raman spectral database of the microorganisms [32–37]. However, many of these studies have been limited to only separate Gram-positive or only Gram-negative bacterial isolates, and have not attempted to evaluate a dataset consisting of both Gram-positive and Gram-negative mixed cultures as well as additional wound components. Studies continue on the characterization and identification of bacteria using SERS and Raman spectroscopy.

1.3 Future Use

The ultimate goal of SERS in bacterial diagnostics is to develop its capability to be used in the clinic with a multimicrobial, multifactorial sample taken directly from the patient. This idea of a point-of-use technique, wherein the sample will be directly analyzed from the wound, blood, or effluent, would save a significant amount of time and may capture pathogens that are difficult or dangerous to culture in the laboratory. Bacterial identification and discrimination of antimicrobial susceptibility is the key to improving patient care and may be vital to reducing and preventing further spread of drug-resistant organisms. A test that has minimal cost, does not destroy the sample, can be used directly from patient samples, and has maximum specificity is likely to revolutionize the bacterial diagnostic world. Although studies are ongoing and SERS shows great promise, current SERS technology has not reached this point and pure cultures are still required for bacterial differentiation.

As SERS, and other technologies, are leveraged and incorporated into the clinical setting, we will be able to move toward the goal of bringing precision medicine to anti-microbial therapy. Currently, we are working within a twentieth century construct of medical diagnosis and delivery, using tools developed in the middle of that century if not earlier. As was highlighted in the recent Institute of Medicine report, *Improving Diagnosis in Health Care* [38], these current tools fall short and need improvement. Rapid evaluation, validation, implementation, and education of such novel approaches is a key step in moving out of the current paradigm toward accurate and timely diagnoses that inform precision medicine approaches to disease states. While all of these steps will require a significant investment of resources, the value gained from such current work on approaches like SERS will ultimately pay off with improved outcomes and lower costs.

1.4 Conclusions

Molecular diagnostics of bacterial pathogens, in particular, spectral-based diagnostics, are one of the fastest growing fields in the clinical laboratory. These molecular tools and procedures are slowly replacing or complementing culture-based, biochemical, and immunological assays in bacterial identification and antimicrobial resistance profiling. Although the current parameters for the use of SERS require a pure culture and are relatively monoparametric, current breakthroughs and testing are pushing the technology to new levels and thus changing the face of modern bacterial diagnostics. It would not be unexpected that within the next few years, physicians will be able to have bacterial identification and resistance profiling occurring directly from multimicrobial, multiparametric patient samples within hours as opposed to days. These developments and new approaches will directly improve patient care and assist in the prevention of the spread of antimicrobial resistance.

2 Materials

2.1 Bacterial Culture Components

1. Lysogeny broth agar (LBA) plates are used to culture bacterial strains in an incubator set at 37 °C.
2. Sterilized 10 μL inoculating loops—for transfers of colonies and bacterial matter.

2.2 Silver Nanoparticle Components

1. Glassware, including conical flask (500 mL) and beakers (100 or 200 mL).
2. Thermometer.
3. Magnetic stirrer.
4. Stirrer hot plate.
5. Aluminum foil.
6. Ice bath.
7. $AgNO_3$ (>99%, Sigma-Aldrich, St. Louis, MO, USA).
8. Trisodium citrate (99.66%, Fisher Scientific, Waltham, MA, USA).
9. Distilled water.
10. Refrigerator (4 °C).
11. 10 mL centrifuge tubes.
12. Syringe and 0.20 μm syringe filters.
13. Centrifuge.
14. Gold slides.
15. Deionized water.
16. Vortex machine.
17. Nanodrop or comparable UV-Vis spectrometer.

2.3 Raman Spectroscopy Components

1. Aluminum foil.
2. Micropipette (capable of dispensing 1–2 μL) and disposable tips.
3. Raman spectroscopic instrument (*see* **Note 1**).

3 Methods

All procedures should be carried out at room temperature unless otherwise specified.

3.1 Growth of Bacterial Culture

Twelve to twenty-four hours *prior* to the collection of SERS spectra, bacterial cultures should be prepared. All samples must be handled in a biosafety hood.

1. From the clinical sample being tested, a single isolated bacterial colony should be picked using a sterile Nunc inoculation loop and used to streak an LBA plate; in the case of a liquid culture/effluent, LBA plates should be streaked to completely cover the entire area of the plate depending on growth patterns and colony density since some bacteria may not grow well in a laboratory environment.
2. Cultures should be incubated for 24 h at 37 °C.

3.2 Silver Nanoparticle Preparation (See Note 1)

1. Prepare an ice bath.
2. Make the silver nitrate solution by suspending 18 mg of $AgNO_3$ in 2 mL distilled water in a small beaker.
3. Prepare 10 mL of 1% w/v trisodium citrate by dissolving 100 mg of trisodium citrate in 9 mL of distilled water in a small beaker. Add remaining distilled water until total volume of solute totals 10 mL.
4. Clean and rinse the 500 mL conical flask, magnetic stirrer, and thermometer thoroughly with detergent and rinse twice with distilled water.
5. Add 100 mL distilled water and magnetic stirrer to a conical flask. Place the flask on the hot plate and suspend the thermometer in the water. Cover the flask opening with aluminum foil to minimize water loss.
6. With continuous stirring, bring the contents of the flask to 45 °C.
7. Add the $AgNO_3$ solution to the flask when the contents of the flask reach 45 °C.
8. Continue to heat and stir the solution to a boiling point.
9. Introduce 2 mL of 1% (w/v) trisodium citrate to the flask.
10. Continue to boil and stir the solution for 10 min. The color should gradually change from colorless to a greenish gray solution (approximately 5 min after having added the trisodium citrate), once the silver nanoparticles have been synthesized.
11. Take the conical flask off the heat and cool rapidly in an ice bath to quench any further reaction.
12. Immediately pipette ~10 mL of the silver nanoparticle solution into centrifuge tubes on ice.
13. Store all centrifuge tubes at 4 °C refrigerator, in a dark box (or covered well with aluminum foil), until needed (*see* **Note 2**).

3.3 Slide Preparation

1. Use a Vortex mixer to thoroughly mix the silver nanoparticle solution prior to filtering.
2. Filter a 10 mL aliquot of silver nanoparticle solution through a 0.20 μm syringe filter two to three times to promote particle size uniformity.
3. Centrifuge the solution at 4000 × *g* for 30 min at 4 °C (*see* **Note 3**).
4. Pipette off the supernatant carefully and discard supernatant.
5. Refill tube with deionized water to the 10 mL mark and use a Vortex to mix the pellet and water.
6. Repeat **steps 3** and **4** one additional time.

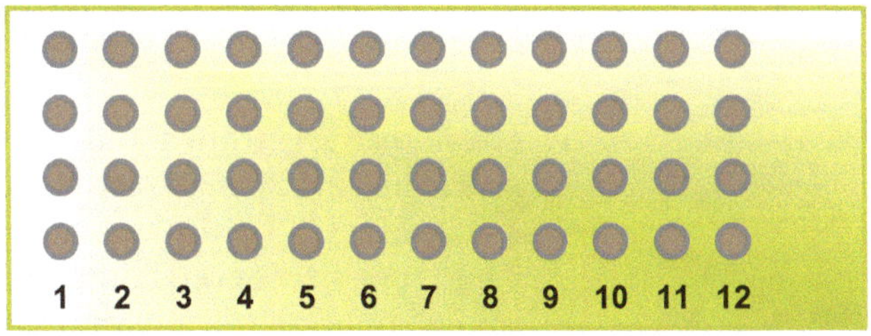

Fig. 3 Schematic of array format of nanoparticle solution deposited onto a gold slide

7. After washing twice, discard the supernatant and transfer the pellet to a new 1.5 mL centrifuge tube.
8. Add deionized water to the pellet to the 1 mL mark. Vortex to mix.
9. Take UV-Vis of the new tube prior to use to determine size, concentration, and uniformity of nanoparticle solution (*see* **Note 4**).
10. After confirming the particles are the desired size, pipette 1.5 μL of the concentrated silver nanoparticle solution directly to the gold slides in an array format (Fig. 3). Allow the spots to dry and deposit an additional 1.5 μL.
11. Allow all spots to dry completely before applying sample (bacteria solution or effluent) to the gold slides (*see* **Note 5**).
12. Pipette 1.5 μL of bacteria solution or effluent directly onto dried spots.
13. To collect traditional Raman spectra of bacteria samples or effluent, skip the silver nanoparticle preparation and pipette 1.5 μL of bacteria solution or effluent directly onto the gold slide in an array format (Fig. 3).

3.4 Spectrum Collection

1. Complete Raman instrument performance verification prior to spectral collection (as per manufacturer's User Manual).
2. Collect Raman spectra and/or SERS spectra using optimized parameters. For example, maximum signal was achieved by acquiring 20 accumulations with 5 s exposures, but enabled total collection times of less than 2 min for each sample. Figure 2 demonstrates an acceptable signal-to-noise ratio for data analysis.
3. Collect at least three replicates per sample for traditional Raman spectra and at least five replicates per sample for SERS spectra.
4. Proceed with appropriate data analysis (*see* **Note 6**).

4 Notes

1. The Raman spectrometer used for this series of experiments was a RamanRxn1™ analyzer (Kaiser Optical Systems Inc., Ann Arbor, MI, USA) with excitation wavelength of 830 nm and a 1 mm spot size. Several different Raman instruments/configurations could be utilized, with success; any Raman analyzer with a small spot size ≤1 mm and visible or near-infrared excitation wavelengths could collect SERS spectra.
2. If the 1 mL aliquot of silver nanoparticle solution is stored overnight for later use, another UV-Vis spectrum must be recorded on each day of use. Once confirmed that the particles have not degraded, pipette out the top 90% of solution and transfer to a new 1.5 mL tube (do not Vortex the solution because particles may have aggregated at the bottom). However, we find that the nanoparticles perform best when prepared the same day as spectral collection.
3. Centrifuge speed may need to be modified based on particle size. If there appears to be dark discoloration on the walls of the tube, the spin speed should be decreased. If the pellet is small and the supernatant is still yellowish in color, the spin speed can be increased.
4. Always take UV-Vis spectrum of the silver nanoparticle solution and record the following before using: peak wavelength, absorbance at peak wavelength, and FWHH (full width at half height). The absorbance and location of the peak wavelength varied some for the silver nanoparticles due to different particles sizes. Absorbance maxima were between 400 and 450 nm, with mean silver nanorod lengths of approximately 150 nm.
5. To make bacteria solution, bacteria grown on LBA plates were transferred into centrifuge tubes and vortexed gently with 5 μL deionized water per mg of bacteria.
6. All spectra were imported into MATLAB® (Natick, MA, USA) for data analysis. Spectra were averaged per sample and truncated to a 600–1800 cm^{-1} spectral window for Raman and SERS of isolates, and a 600–1600 cm^{-1} spectral window for SERS of effluent. Spectra were preprocessed by applying baseline removal and intensity normalization along with mean centering. Using the preprocessed data, PLSDA models were developed to discriminate between Gram-positive and Gram-negative bacteria for isolate data, and between uninfected, Gram-positive and Gram-negative infected samples for effluent spectra. The compiled PLSDA prediction values were used to create a HCA model with Euclidean distance metric and Ward's linkage (Fig. 4).

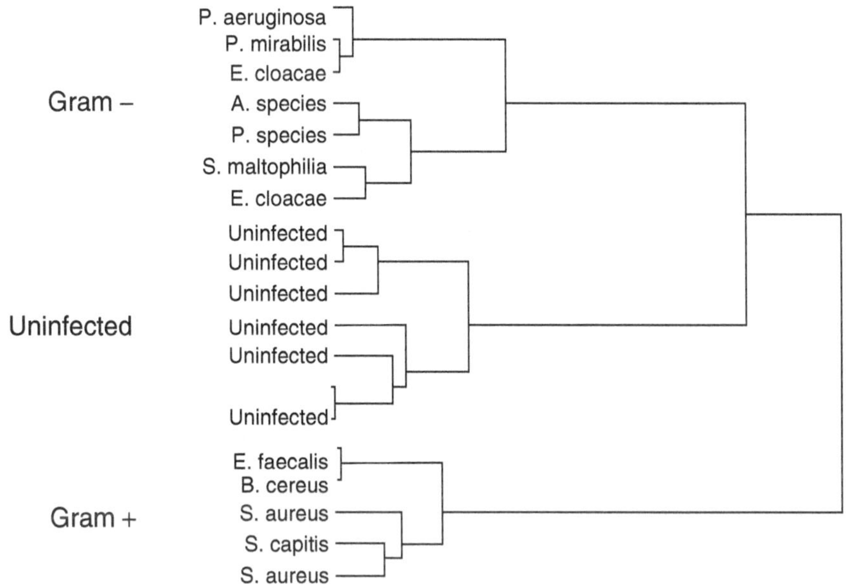

Fig. 4 Hierarchical clusters generated using SERS spectra of 19 effluent samples modeled to classify effluent from uninfected wounds and those infected by Gram-positive and Gram-negative bacteria

Acknowledgments

This work was prepared as part of the authors' official duties. Title 17 U.S.C. §105 provides that "Copyright protection under this title is not available for any work of the United States Government." Title 17 U.S.C. §101 defines a U.S. Government work as a work prepared by a military service member or employee of the U.S. Government as part of that person's official duties. The views expressed in this article are those of the author and do not necessarily reflect the official policy or position of the Department of the Navy, Department of Defense, nor the U.S. Government. This effort was supported (in part) by the U.S. Navy Bureau of Medicine and Surgery under the Medical Development Program and Office of Naval Research work unit number (602115HP.3720.001.A1015) and USAMRAA award W81XWH-13-2-0039. This study was performed under MUA 228 with the Walter Reed Army Institute of Research. I/We certify that all individuals who qualify as authors have been listed; each has participated in the conception and design of this work, the analysis of data (when applicable), the writing of the document, and the approval of the submission of this version; that the document represents valid work; that if we used information derived from another source, we obtained all necessary approvals to use it and made appropriate acknowledgements in the document; and that each takes public responsibility for it.

References

1. Livermore DM, Wain J (2013) Revolutionising bacteriology to improve treatment outcomes and antibiotic stewardship. Infect Chemother 45:1–10
2. Kootallur BN, Thangavelu CP, Mani M (2011) Bacterial identification in the diagnostic laboratory: how much is enough? Indian J Med Microbiol 29:336–340
3. Wills H et al (2009) Raman spectroscopy detects and distinguishes neuroblastoma and related tissues in fresh and (banked) frozen specimens. J Pediatr Surg 44:386–391
4. Harvey TJ et al (2008) Spectral discrimination of live prostate and bladder cancer cell lines using Raman optical tweezers. J Biomed Opt 13:064004
5. Andrade PO et al (2007) Study of normal colorectal tissue by FT-Raman spectroscopy. Anal Bioanal Chem 387:1643–1648
6. Buschman HP et al (2001) Raman microspectroscopy of human coronary atherosclerosis: biochemical assessment of cellular and extracellular morphologic structures in situ. Cardiovasc Pathol 10:69–82
7. Carden A, Rajachar RM, Morris MD, Kohn DH (2003) Ultrastructural changes accompanying the mechanical deformation of bone tissue: a Raman imaging study. Calcif Tissue Int 72:166–175
8. Chan KL et al (2008) A coordinated approach to cutaneous wound healing: vibrational microscopy and molecular biology. J Cell Mol Med 12(5b):2145–2154
9. Chowdary MV et al (2007) Discrimination of normal and malignant mucosal tissues of the colon by Raman spectroscopy. Photomed Laser Surg 25:269–274
10. Crane NJ, Popescu V, Morris MD, Steenhuis P, Ignelzi MA Jr (2006) Raman spectroscopic evidence for octacalcium phosphate and other transient mineral species deposited during intramembraneous mineralization. Bone 39:434–442
11. Haka AS et al (2006) In vivo margin assessment during partial mastectomy breast surgery using Raman spectroscopy. Cancer Res 66:3317–3322
12. Jess PR et al (2007) Early detection of cervical neoplasia by Raman spectroscopy. Int J Cancer 121:2723–2728
13. Koljenovic S et al (2007) Raman spectroscopic characterization of porcine brain tissue using a single fiber-optic probe. Anal Chem 79:557–564
14. Leroy G, Penel G, Leroy N, Brès E (2002) Human tooth ename: a Raman polarized approach. Appl Spectrosc 56:1030–1034
15. McGill N, Dieppe PA, Bowden M, Gardiner DJ, Hall M (1991) Identification of pathological mineral deposits by Raman microscopy. Lancet 337:77–78
16. Robichaux-Viehoever A et al (2007) Characterization of Raman spectra measured in vivo for the detection of cervical dysplasia. Appl Spectrosc 61:986–993
17. Shetty G, Kendall C, Shepherd N, Stone N, Barr H (2006) Raman spectroscopy: elucidation of biochemical changes in carcinogenesis of oesophagus. Br J Cancer 94:1460–1464
18. Shim MG, Wilson BC, Marple E, Wach M (1999) Study of fiber-optic probes for *in vivo* medical Raman spectroscopy. Appl Spectrosc 53:619–627
19. Wang TD, Van Dam J (2004) Optical biopsy: a new frontier in endoscopic detection and diagnosis. Clin Gastroenterol Hepatol 2:744–753
20. LaPlant F (2010) In: Matousek P, Morris MD (eds) Emerging Raman applications and techniques in biomedical and pharmaceutical fields. Springer-Verlag, Berlin, pp 3–4
21. Le Ru EC, Etchegoin PG (2009) Principles of surface-enhanced Raman spectroscopy and related plasmonic effects. Elsevier, Oxford
22. Fleischmann M, Hendra PJ, McQuillan AJ (1974) Raman spectra of pyridine adsorbed at a silver electrode. Chem Phys Lett 26(2):163–166
23. Buijtels PC et al (2008) Rapid identification of mycobacteria by Raman spectroscopy. J Clin Microbiol 46:961–965
24. Escoriza MF, VanBriesen JM, Stewart S, Maier J, Treado PJ (2006) Raman spectroscopy and chemical imaging for quantification of filtered waterborne bacteria. J Microbiol Methods 66:63–72
25. Maquelin K et al (2002) Identification of medically relevant microorganisms by vibrational spectroscopy. J Microbiol Methods 51:255–271
26. Wu Q et al (2000) Intensities of *E. coli* nucleic acid Raman spectra excited selectively from whole cells with 251-nm light. Anal Chem 72:2981–2986
27. Zeiri L, Bronk BV, Shabtai Y, Eichler J, Efrima S (2004) Surface-enhanced Raman spectroscopy as a tool for probing specific biochemical components in bacteria. Appl Spectrosc 58:33–40

28. Maquelin K, Dijkshoorn L, van der Reijden TJ, Puppels GJ (2006) Rapid epidemiological analysis of Acinetobacter strains by Raman spectroscopy. J Microbiol Methods 64:126–131
29. Maquelin K et al (2000) Raman spectroscopic method for identification of clinically relevant microorganisms growing on solid culture medium. Anal Chem 72:12–19
30. Kalasinsky KS et al (2007) Raman chemical imaging spectroscopy reagentless detection and identification of pathogens: signature development and evaluation. Anal Chem 79:2658–2673
31. Zeiri L, Bronk BV, Shabtai Y, Czege J, Efrima S (2002) Silver metal induced surface enhanced Raman of bacteria. Colloids Surf A Physicochem Eng Asp 208:357–362
32. Ghebremedhin M, Yesupriya S, Luka J, Crane NJ (2015) Validation of hierarchical cluster analysis for identification of bacterial species using 42 bacterial isolates. Proc SPIE 9318:93180W
33. Premasiri WR, Sauer-Budge AF, Lee JC, Klapperich CM, Ziegler LD (2012) Rapid bacterial diagnostics via surface enhanced Raman microscopy. Spectroscopy (Springf) 27:s8–s31
34. Ozaki Y, Kneipp K, Aroca R (2014) Frontiers of surface-enhanced Raman scattering: single nanoparticles and single cells. Wiley, West Sussex
35. Rosch P et al (2005) Chemotaxonomic identification of single bacteria by micro-Raman spectroscopy: application to clean-room-relevant biological contaminations. Appl Environ Microbiol 71:1626–1637
36. Premasiri WR, Moir DT, Klempner MS, Zeigler LD (2007) In: Kneipp K, Aroca R, Kneupp H, Wentrup-Byrne E (eds) New approaches in biochemical spectroscopy. Oxford University Press, New York, p 164
37. Crane NJ, Elster EA (2012) Profiling wound healing with wound effluent: Raman spectroscopic indicators of infection. Proc SPIE 8220:82200S
38. National Academies of Sciences, Engineering, and Medicine (2015) Improving diagnosis in health care. The National Academies Press, Washington, DC

Index

Kimberly A. Bishop-Lilly (ed.), *Diagnostic Bacteriology: Methods and Protocols*, Methods in Molecular Biology, vol. 1616, DOI 10.1007/978-1-4939-7037-7, © Springer Science+Business Media LLC 2017

Kolibri 360 GmbH
Ferdinand-Jühlke-Straße 7
99095 Erfurt, Deutschland
produktsicherheit@kolibri360.de